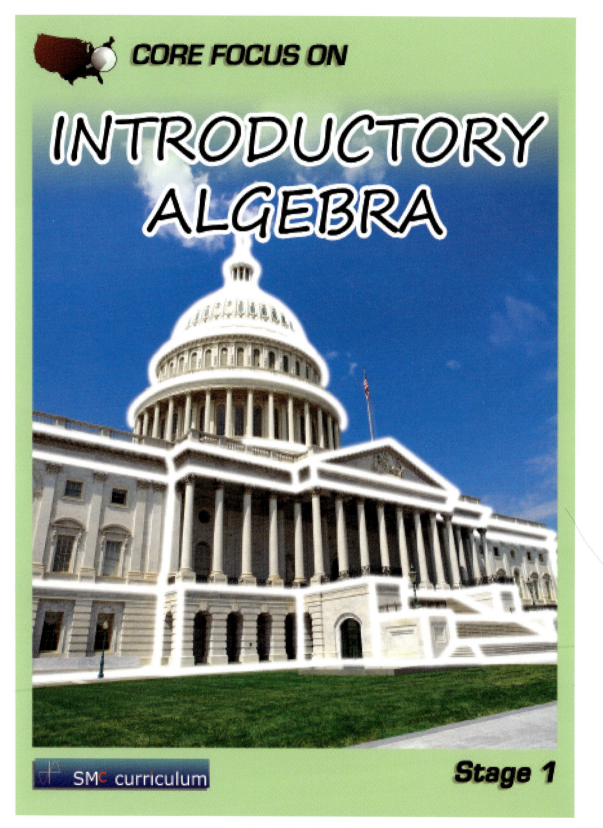

CORE FOCUS ON

INTRODUCTORY ALGEBRA

Stage 1

SMc curriculum

AUTHORS

Shannon McCaw

Beth Armstrong • Matt McCaw • Sarah Schuhl • Michelle Terry • Scott Valway

COVER PHOTOGRAPH

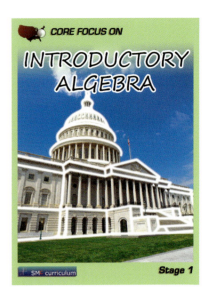

U.S. Capitol
Located in Washington DC, the U.S. Capitol is
where the United States Congress meets. Originally
constructed beginning in 1793, the U.S. Capitol
has become a centerpiece of the National Mall and
a symbol of the United States government.
©*iStockphoto.com/SOMATUSCANI*

ISBN: 978-1-938801-72-3

4 5 6 7 8 9 10

ABOUT THE AUTHORS

From left to right: Beth Armstrong, Matt McCaw, Shannon McCaw, Scott Valway, Michelle Terry, Sarah Schuhl

SERIES AUTHOR

Shannon McCaw has been a classroom teacher in the Newberg and Parkrose School Districts in Oregon. She has been trained in Professional Learning Communities, Differentiated Instruction and Critical Friends. Shannon currently works as a consultant with math teachers from over 100 districts around Oregon. Shannon's areas of expertise include the Common Core State Standards, curriculum alignment, assessment best practices and instructional strategies. She has a degree in Mathematics from George Fox University and a Masters of Arts in Secondary Math Education from Colorado College.

CONTRIBUTING AUTHORS & EDITORS

Beth Armstrong has been a classroom teacher in the Beaverton School District in Oregon. She has received training in Talented and Gifted Instruction. She has a Masters in Curriculum and Instruction from Washington State University.

Matt McCaw has been a classroom teacher, math/science TOSA and special education case-manager in several Oregon school districts. Matt has most recently worked as a curriculum developer and math coach for grades 6-8. He is trained in Differentiated Instruction, Professional Learning Communities, Critical Friends Groups and Understanding Poverty. Matt has a Masters of Special Education from Western Oregon University.

Sarah Schuhl has been a classroom teacher, secondary math instructional coach and district-wide K-12 math instructional specialist, most recently in the Centennial School District in Oregon. Sarah currently works as a Solution Tree associate and an educational consultant supporting and challenging teachers in the areas of math instruction and alignment to the Common Core State Standards, common assessments for all subjects and grade levels and professional learning communities. From 2010–2013, Sarah served as a member and chair of the National Council of Teachers of Mathematics editorial panel for their Mathematics Teacher journal. Sarah earned a Masters of Science in Teaching Mathematics from Portland State University.

Michelle Terry has been a classroom teacher in the Estacada and Newberg School Districts in Oregon. Michelle has received training in Professional Learning Communities, Critical Friends, ELL Instructional Strategies, Proficiency-Based Grading and Lesson Design, Power Strategies for Effective Teaching, and Classroom Love and Logic. Michelle has an Interdisciplinary Masters from Western Oregon University. She currently teaches mathematics at Newberg High School.

Scott Valway has been a classroom teacher in the Tigard-Tualatin, Newberg and Parkrose School Districts in Oregon. Scott has been trained in Differentiated Instruction, Professional Learning Communities, Critical Friends, Discovering Algebra, Pre-Advanced Placement, Assessment Writing and Credit by Proficiency. Scott has a Masters of Science in Teaching from Oregon State University. He currently teaches math at Parkrose High School.

COMMON CORE STATE STANDARDS

Grade 6 Overview

The complete set of Common Core State Standards can be found at http://www.corestandards.org/. This book focuses on the highlighted Common Core State Standards shown below.

Ratios and Proportional Relationships

- Understand ratio concepts and use ratio reasoning to solve problems.

The Number System

- Apply and extend previous understanding of multiplication and division to divide fractions by fractions.

- Compute fluently with multi-digit numbers and find common factors and multiples.

- Apply and extend previous understandings of numbers to the system of rational numbers.

Expressions and Equations

- Apply and extend previous understandings of arithmetic to algebraic expressions.

- Reason about and solve one-variable equations and inequalities.

- Represent and analyze quantitative relationships between dependent and independent variables.

Geometry

- Solve real-world and mathematical problems involving area, surface area and volume.

Statistics and Probability

- Develop understanding of statistical variability.

- Summarize and describe distributions.

Mathematical Practices

1. Make sense of problems and persevere in solving them.

2. Reason abstractly and quantitatively.

3. Construct viable arguments and critique the reasoning of others.

4. Model with mathematics.

5. Use appropriate tools strategically.

6. Attend to precision.

7. Look for and make use of structure.

8. Look for and express regularity in repeated reasoning.

CORE FOCUS ON INTRODUCTORY ALGEBRA

CONTENTS IN BRIEF

CORE FOCUS ON INTRODUCTORY ALGEBRA

BLOCK 1 ~ ORDER OF OPERATIONS

BLOCK 2 ~ ALGEBRAIC EXPRESSIONS

HOW TO USE YOUR MATH BOOK

Your math book has features that will help you be successful in this course. Use this guide to help you understand how to use this book.

Lesson Target

 Look in this box at the beginning of every lesson to know what you will be learning about in each lesson.

Vocabulary

Each new vocabulary word is printed in red. The definition can be found with the word. You can also find the definition of the word in the glossary which is in the back of this book.

Explore!

Some lessons have **EXPLORE!** activities which allow you to discover mathematical concepts. Look for these activities in the Table of Contents and in lessons next to the purple line.

Examples

Examples are useful because they remind you how to work through different types of problems. Look for the word **EXAMPLE** and the green line.

Helpful Hints

Helpful hints and important things to remember can be found in green callout boxes.

Blue Boxes

A blue box holds important information or a process that will be used in that lesson. Not every lesson has a blue box.

 This calculator icon will appear in Lessons and Exercises where a calculator is needed. Your teacher may want you to use your calculator at other times, too. If you are unsure, make sure to ask if it is the right time to use it.

EXERCISES

The **EXERCISES** are a place for you to find practice problems to determine if you understand the lesson's target. You can find selected answers in the back of this book so you can check your progress.

REVIEW

The **REVIEW** provides a set of problems for you to practice concepts you have already learned in this book. The **REVIEW** follows the **EXERCISES** in each lesson. There is also a **REVIEW** section at the end of each Block.

TIC-TAC-TOE ACTIVITIES

Each Block has a Tic-Tac-Toe board at the beginning with activities that extend beyond the Common Core State Standards. The Tic-Tac-Toe activities described on the board can be found throughout each Block in yellow boxes.

CAREER FOCUS

At the end of each Block, you will find an autobiography of an individual. Each one describes what they like about their job and how math is used in their career.

CORE FOCUS ON MATH
STAGE 1

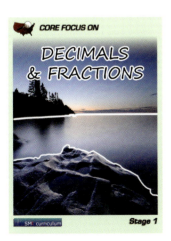

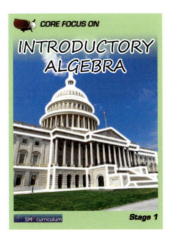

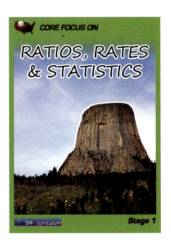

CORE FOCUS ON INTRODUCTORY ALGEBRA

BLOCK 1 ~ ORDER OF OPERATIONS

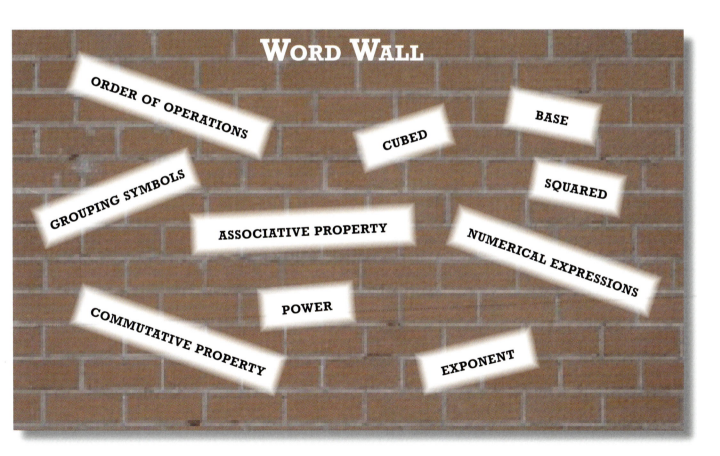

BLOCK 1 ~ ORDER OF OPERATIONS
TIC-TAC-TOE

HALF-LIFE

Investigate the half-life of a chemical. Create a diagram to illustrate what occurs to the chemical over time.

See page 17 for details.

PERFECT SQUARES

Examine perfect squares up to 400. Make a prediction about the possible last digits of perfect squares.

See page 10 for details.

ORDER OF OPERATIONS POETRY

Write two different styles of poems about the order of operations in mathematics.

See page 22 for details.

PROPERTIES POSTER

Create a poster that will help your classmates remember the Commutative and Associative Properties.

See page 22 for details.

100 PUZZLE

Use all positive single-digit whole numbers to create numerical expressions which equal 100.

See page 13 for details.

DISTANCES TO PLANETS

Research the distance to the planets in our solar system. Express very large distances in scientific notation.

See page 25 for details.

FRACTALS

Examine Sierpiñski's Triangle and predict how many upward-pointing triangles can be found in each stage.

See page 10 for details.

GRID GAME

Create a grid pattern game using numerical expressions that include the four operations (+, −, ×, ÷).

See page 5 for details.

NUMBER TRICK

Try a number trick. Describe someone else's number trick and then make your own.

See page 17 for details.

THE FOUR OPERATIONS

 Find values of expressions involving addition, subtraction, multiplication and division.

Kirby and Angelea were asked to find the value of the math problem shown below.

$$4 + 3 \times 2 - 1$$

They disagreed on the answer. Angelea believed the answer was 13. Kirby thought the answer was 9. Who do you agree with?

A **numerical expression** is a combination of numbers and operations. Many numerical expressions contain more than one operation. In the problem above, the students chose two different orders to perform the operations.

- Angelea did each operation from left to right.

  $4 + 3 \times 2 - 1$
  $= 7 \times 2 - 1$
  $= 14 - 1$
  $= 13$

- Kirby multiplied first. Then he added and subtracted from left to right.

  $4 + 3 \times 2 - 1$
  $= 4 + 6 - 1$
  $= 10 - 1$
  $= 9$

Mathematicians have established an **order of operations**. These describe the rules to follow when evaluating an expression with more than one operation. This way everyone gets the same solution for a numerical expression. <u>KIRBY</u> followed the correct order of operations.

> ### ORDER OF OPERATIONS
> 1. Multiply and divide from left to right.
> 2. Add and subtract from left to right.

EXAMPLE 1 | **Find the value of $5 \times 8 - 12 \div 3$.**

SOLUTION | Multiply and divide from left to right. | $5 \times 8 - 12 \div 3 = 40 - 4$
| Subtract. | $40 - 4 = 36$
| $5 \times 8 - 12 \div 3 = 36$

EXAMPLE 2 | **Find the value of $20 + 16 \div 4 - 2 \times 9$.**

SOLUTION | Multiply and divide from left to right. | $20 + 16 \div 4 - 2 \times 9 = 20 + 4 - 18$
| Add and subtract from left to right. | $20 + 4 - 18 = 6$
| $20 + 16 \div 4 - 2 \times 9 = 6$

EXAMPLE 3

Saul's mom and dad took him to a basketball game for his birthday. He was allowed to bring four friends with him. His parents bought $24 tickets for each of the five boys so they could sit close to the court. The parents' tickets cost $10 each. Find the total cost of the game tickets.

SOLUTION

Write the problem.	$24 × 5 + $10 × 2
Multiply to find the total cost for the boys' tickets.	$24 × 5 = $120
Multiply to find the cost for the parents' tickets.	$10 × 2 = $20
Add the cost for the boys' tickets and the cost for the parents' tickets.	$120 + $20 = $140

It will cost $140 for the group to go to the basketball game.

EXERCISES

Find the value of each expression.

1. $7 + 2 \times 5$

2. $12 \div 2 - 3 + 1$

3. $15 - 10 \div 5$

4. $4 \times 6 + 7 \times 3$

5. $20 \div 5 \times 4$

6. $12 - 10 + 40 \div 8$

7. $6 \times 10 + 22 \div 2$

8. $9 - 4 \times 2 - 1$

9. $100 + 3 \times 20 - 15$

10. $5.7 + 2 \times 4.3$

11. $4 \times 3 + 5 \times 2 - 3 \times 3$

12. $36 \div 6 \times 2 + 4$

13. $75 \div 3 + 5 \times 5$

14. $2 \div 2 + 8 - 3$

15. $5.5 \div 1.1 + 2.6$

16. Joaquin made errors in each of the problems below. Find and describe each error. Find the correct solutions.

a. $25 - 6 \times 2 + 1$
$= 19 \times 2 + 1$
$= 38 + 1$
$= 39$

b. $7 + 12 \div 2 \times 3$
$= 7 + 12 \div 6$
$= 7 + 2$
$= 9$

Copy each expression and insert one of the four operations (+, −, ×, ÷) in each box so the numerical expression equals the stated amount. Use mathematics to show that your answer is correct.

17. $6 \,\square\, 3 \,\square\, 5 = 21$

18. $28 \,\square\, 7 \,\square\, 1 = 3$

19. $10 \,\square\, 4 \,\square\, 5 \,\square\, 2 = 4$

20. Four friends went out to dinner. The taxi ride to the restaurant cost $18. Each one ordered the all-you-can-eat buffet for $12. What was the total cost for the evening? Use words and/or numbers to show how you determined your answer.

21. Terry went to the store and bought 3 bags of chips for $2.50 each. She also bought 2 boxes of juice for $1.25 each. What was the total cost of the food and beverages purchased by Terry? Use mathematics to justify your answer.

22. The official state dance in Washington is the square dance. The Friday Night Square Dance-Off at a local dance club charges $7 per adult and $4 per child. What is the total admission cost for a group of 3 adults and 6 children?

23. Use all of the following numbers and any of the four operations (+, −, ×, ÷) to write a numerical expression that equals the amount given. Use mathematics to show that your answer is correct. You can use the same operation more than once in a numerical expression.

<h2 style="text-align:center">5, 2, 3, 4</h2>

    a. = 14   **b.** = 7   **c.** = 30   **d.** = 0

TIC-TAC-TOE ~ GRID GAME

Create a grid game that will help other students review the order of operations using the four operations (+, −, ×, ÷). On a sheet of white paper, create an 8 inch by 8 inch grid that is divided into 2 inch squares. Create a numerical expression that includes multiple operations and write this expression on the inside edge of a small square. Use the order of operations to find the value of the expression. Insert the answer on the adjacent square as seen below.

		$2 \times 5 + 1$	
		11	

Continue creating numerical expressions that require the use of the order of operations. Put the corresponding answers on the adjacent square. Every side of each square should have either an expression or the value of an expression except the outside edges of the grid which can be left blank. You may create distractors on the outside edges of the grid that do not have corresponding answers. Once your grid game is complete, make a copy to turn in. Cut the pieces of the original apart. Ask a friend to try to put the puzzle back together.

POWERS AND EXPONENTS

LESSON 1.2

Write and compute expressions with powers.

Kirkland School District set up a phone tree for their teachers in case snow caused school to be cancelled. A principal called three teachers to tell them school was cancelled. Each of those three teachers called three more teachers. Whenever a teacher heard the news, that teacher called an additional three teachers. This continued until all teachers had been called.

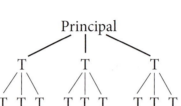

Round	Number of Teachers Who Have Been Contacted During Each Round	Numerical Expression
1	3	3
2	9	3 × 3
3	27	3 × 3 × 3
4	81	3 × 3 × 3 × 3

When a numerical expression is the product of a repeated factor, it can be written using a **power**. A power consists of two parts, the **base** and the **exponent**. The base of the power is the repeated factor. The exponent shows the number of times the factor is repeated.

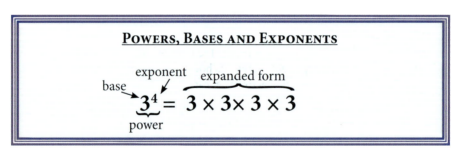

POWERS, BASES AND EXPONENTS

$$3^4 = 3 \times 3 \times 3 \times 3$$

It is important to know how to read powers correctly.

Power	Reading the Expression	Expanded Form	Value
5^2	"five to the second power" or "five squared"	5 × 5	25
6^3	"six to the third power" or "six cubed"	6 × 6 × 6	216
2^4	"two to the fourth power"	2 × 2 × 2 × 2	16

Step 1: Copy the table below.

Number of Folds	Total Number of Rectangles	Expanded Form	Power
1			
2			
3			
4			
5			

Step 2: Take an $8\frac{1}{2}$" by 11" sheet of notebook paper. Fold it in half and then open it up. Count the number of rectangles. Record it in the table. Do not fill in the shaded boxes.

Step 3: Repeat the first fold and then fold the paper in half again. Open it up and count the number of rectangles. Record it in the table. Write this expression in expanded form and write the power. You are doubling the number of rectangles each time.

Step 4: Repeat the process in **Step 3** until you have completed five folds.

Step 5: After a while the paper becomes too hard to fold. Can you predict how many rectangles would be formed if the paper was folded in half 9 times? Write your answer in expanded form and as a power. Calculate the value.

EXAMPLE 1

Write the numerical expression as a power.
a. $6 \times 6 \times 6 \times 6$ **b.** 4×4 **c.** $9 \times 9 \times 9 \times 9 \times 9 \times 9 \times 9$

SOLUTIONS

a. 6^4 **b.** 4^2 **c.** 9^7

Remember that the base is the repeated factor. The exponent represents how many times the factor is repeated.

EXAMPLE 2

Write each power in expanded form and find the value.
a. 3^2 **b.** 1^5 **c.** 4^3

SOLUTIONS

a. Expanded Form: 3×3
Value: 9

b. Expanded Form: $1 \times 1 \times 1 \times 1 \times 1$
Value: 1

c. Expanded Form: $4 \times 4 \times 4$
Value: 64

EXERCISES

1. Use the power 2^3. Which number represents the:
   a. base?
   b. exponent?
   c. value of the power?

Write the numerical expression as a power.

2. $3 \times 3 \times 3 \times 3 \times 3$

3. $1 \times 1 \times 1 \times 1$

4. 7×7

5. $2 \times 2 \times 2 \times 2 \times 2 \times 2 \times 2$

6. $8 \times 8 \times 8$

7. $11 \times 11 \times 11$

Write each power in expanded form. Find its value.

8. 6^2

9. 1^7

10. 2^3

11. 5^3

12. 3^4

13. 10^2

14. $\left(\frac{1}{2}\right)^2$

15. $\left(\frac{1}{3}\right)^3$

16. 7^1

17. Is 5^3 the same as 3^5? Explain your answer.

Determine which power has the greater value.

18. 2^2 or 1^6

19. 5^2 or 2^5

20. $\left(\frac{1}{2}\right)^2$ or $\left(\frac{1}{5}\right)^2$

21. List the following from least to greatest: $2^5, 3^3, 4^2$. Use mathematics to show that your answer is correct.

22. A banana split has approximately 8^3 calories. Determine the number of calories as a whole number.

23. The volume of a cube can be calculated using the formula *length* × *width* × *height*.
   a. Write the volume of the cube shown in expanded form.
   b. Write the volume of the cube as a power.
   c. Find the value of the volume of the cube.

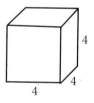

24. Calculators have three different buttons that allow you to find the value of a power. You can use any of the following buttons:

$$\boxed{x^y} \quad \boxed{y^x} \quad \boxed{\wedge}$$

   a. Find the value of 4^7 on a calculator by entering the base. Press one of the buttons shown and then enter the exponent.
   b. Use a calculator to find the value of 3^8.
   c. Use a calculator to find the value of 2^{10}.

25. Tadashi made a deal with his father. If he completed his monthly chores he would get $4 at the end of the first month. If he continued to do his chores, his dad would multiply his pay by four for the second month. His father would continue to multiply his new pay by four as long as Tadashi completed all of his chores each month.

    a. Copy and complete the table. Do not fill in the shaded boxes.

Months Chores Have Been Completed	Pay for Completing Chores	Expanded Form	Power
1			
2			
3			
4			
5			

    b. Do you think this is a reasonable agreement between Tadashi and his dad? Explain your reasoning.

26. Jake made a common mistake when he stated that 2^3 is 6. What did he do wrong? What should he have done to find the correct value of the power?

27. Michelle says that $3^2 \times 3^5$ is the same as 3^7. Do you agree or disagree? Explain your reasoning.

REVIEW

Find the value of each expression.

28. $8 \div 1 \times 3$

29. $24 \div 6 - 2 + 1$

30. $31 - 20 \div 4$

31. $7 \times 10 + 5 \times 11$

32. $2 \times 10 \div 2$

33. $12 - 8 + 32 \div 8$

34. $13 \times 10 + 77 \div 11$

35. $12 - 6 \times 2 + 1$

36. $2.5 + 3.5 \times 1.5$

Copy and insert one of the four operations ($+, -, \times, \div$) in each box so that the numerical expression equals the stated amount. Use mathematics to show that your answer is correct.

37. $27 \boxed{} 9 \boxed{} 3 = 0$

38. $8 \boxed{} 6 \boxed{} 2 = 50$

Tic-Tac-Toe ~ Fractals

A fractal is a geometric shape that is subdivided into parts to create a pattern. Waclaw Sierpiński described Sierpiński's Triangle in 1915. Follow the steps below to create the first five stages of Sierpiński's Triangle.

Stage 1: Draw an equilateral triangle.

Stage 2: Copy your **Stage 1 Triangle**. Connect the midpoints (middle) of each side of the triangle to create a new triangle.

Stage 3: Copy your **Stage 2 Triangle**. Connect the midpoints of each side of the upward pointing triangles to create new triangles.

Stage 4: Repeat **Stage 3** with your **Stage 3 Triangle**.

Stage 5: Repeat **Stage 3** with your **Stage 4 Triangle**.

The number of upward-pointing triangles grows with each stage of the Sierpiński Triangle. Copy the table below and record the number of new upward-pointing triangles at each stage. Can you write the number of upward-pointing triangles as powers? Predict how many new upward-pointing triangles would be found in the tenth stage of Sierpiński's Triangle.

Stage	New Upward-Pointing Triangles	Written as a Power
1		
2		
3		
4		
5		
10		

Tic-Tac-Toe ~ Perfect Squares

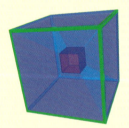

A perfect square is the square of a whole number. For example, the first four perfect squares are:

$$1^2 = 1$$
$$2^2 = 4$$
$$3^2 = 9$$
$$4^2 = 16$$

Step 1: Generate a list of all perfect squares to $20^2 = 400$.

Step 2: There are only six possibilities for the last digit of a perfect square. What are the six possible digits based on your findings in **Step 1**?

Step 3: Make predictions about the relationship of the last digit of the number being squared to the last digit of the perfect square. Summarize your findings.

ORDER OF OPERATIONS WITH POWERS

 Find values of expressions with exponents using the order of operations.

Numerical expressions often include powers. You learned in **Lesson 1.1** that there is a pre-determined order in which you must complete operations (order of operations) to find the value of an expression. When powers are included in an expression, you must find the value of all powers before multiplying, dividing, adding or subtracting.

> ### ORDER OF OPERATIONS
> 1. Find the value of all powers.
> 2. Multiply and divide from left to right.
> 3. Add and subtract from left to right.

When finding the value of a numerical expression using the order of operations, it is very important to keep your work organized. Work through each step one at a time. Show your work line by line. The example shown here is a great way to organize your work.

> **Find the value of**
> $$5 + 3^3 \div 9 + 2$$
>
> Line 1 $= 5 + 27 \div 9 + 2$
> Line 2 $= 5 + 3 + 2$
> Line 3 $= 10$

EXAMPLE 1

Find the value of each expression.

a. $5 + 2^3 \times 3 - 10$ **b.** $4^2 + 5 \div 1^4 \times 7$

$$\boxed{2^3 = 2 \times 2 \times 2 = 8}$$

SOLUTIONS

a. Find the value of all powers.

$5 + 2^3 \times 3 - 10 = 5 + 8 \times 3 - 10$

Multiply.

$5 + 8 \times 3 - 10 = 5 + 24 - 10$

Add and subtract from left to right.
$5 + 2^3 \times 3 - 10 = 19$

$5 + 24 - 10 = 19$

b. Find the value of all powers.

$4^2 + 5 \div 1^4 \times 7 = 16 + 5 \div 1 \times 7$

Multiply and divide from left to right.

$16 + 5 \div 1 \times 7 = 16 + 35$

Add.
$4^2 + 5 \div 1^4 \times 7 = 51$

$16 + 35 = 51$

EXAMPLE 2

The diameter of Saturn is approximately 42^3 miles. The diameter of Neptune is approximately 175^2 miles.
a. Which planet is bigger?
b. What is the difference in the diameter of the two planets?

SOLUTIONS

a. Find the value of the diameter of each planet by writing the expanded form and then multiplying.

<u>Saturn</u>   <u>Neptune</u>
$42 \times 42 \times 42$ 175×175
$= 74{,}088$ $= 30{,}625$
Saturn is bigger.

b. Subtract to find how much larger Saturn's diameter is than Neptune's diameter.

$$74{,}088$$
$$-\ \underline{30{,}625}$$
$$43{,}463 \text{ miles}$$

Saturn's diameter is 43,463 miles bigger than Neptune's diameter.

Notice that you can not determine how much larger the diameter of one planet is than the diameter of the other planet unless you find the values of the powers first.

EXERCISES

Find the value of each expression.

1. $3 + 2^3 \times 4$

2. $15 \div 3 + 3^2 - 9$

3. $30 - 1^4 \times 6$

4. $10^2 \div 2 + 10$

5. $27 \div 3^3 \times 4$

6. $2^4 - 1 + 4^2$

7. $5^2 \times 3 \div 6 \times 3^2$

8. $9^2 + 2^3 + 1$

9. $5 \times 3^2 - 3 \times 2$

10. $3 + 2^5 \times 2$

11. $100 - 3^2 \times 3^2$

12. $6^2 \div 6 \times 2 + 4$

13. $4 \times 5^3 \div 2$

14. $2^4 \div 2^2 - 4$

15. $5 \div 1 + 8^2$

One number in each numerical equation should be squared (have a power of two) so that it equals the stated amount. Rewrite each equation with the appropriate number squared. Use mathematics to show that your answer is correct.

16. $3 \times 5 + 7 = 82$

17. $10 - 4 \div 2 = 9$

18. $4 \times 3 \times 2 + 1 = 97$

19. Exterminators discovered that a certain house was infested with 6^3 spiders. They also found 5^4 ants in the same house.

    a. Which insect population was greater in this house?

    b. How much larger is the population of one insect than the other?

20. An error was made in the order of operations in each problem. Determine on which line each error occurs. Find the correct solution.

    a. $4 + 2^3 \div 2 + 5$ **b.** $10^2 \div 2 \times 5 + 1^3$

          Line 1 $= 6^3 \div 2 + 5$ Line 1 $= 100 \div 2 \times 5 + 1$

          Line 2 $= 216 \div 2 + 5$ Line 2 $= 100 \div 10 + 1$

          Line 3 $= 108 + 5$ Line 3 $= 10 + 1$

          Line 4 $= 113$ Line 4 $= 11$

21. Jennifer said she did not think it mattered if you did operations from left to right or from right to left when finding the value of an expression. Do you agree or disagree? Support your answer with words and/or symbols.

22. Edgar stated that he thought powers could be done from left to right at the same time as multiplication and division because powers are just repeated multiplication. Do you agree or disagree? Support your answer with words and/or symbols.

REVIEW

Write the numerical expression as a power.

23. $2 \times 2 \times 2 \times 2 \times 2$ **24.** $9 \times 9 \times 9$ **25.** $12 \times 12 \times 12 \times 12 \times 12 \times 12$

Determine which power has the greater value.

26. 3^2 or 2^3 **27.** 10^2 or 5^3 **28.** 6^3 or 3^5

29. List the following from least to greatest: $6^2, 2^6, 3^3$.

30. List the following from least to greatest: $4^3, 1^{70}, 9^2$.

 **TIC-TAC-TOE ~ 100 PUZZLE**

Use the digits 1, 2, 3, 4, 5, 6, 7, 8, 9 in this order and any combination of the operations ($+, -, \times, \div$) to write expressions equal to 100.

For example: $1 \times 2 \times 3 - 4 + 5 + 6 + 78 + 9 = 100$

Find four or more expressions. There are at least 20.

ORDER OF OPERATIONS WITH GROUPING SYMBOLS

LESSON 1.4

 Find values of expressions with grouping symbols using the order of operations.

Y ou have learned the order of operations for numerical expressions including powers, multiplication, division, addition and subtraction. Grouping symbols group parts of an expression. Grouping symbols must be completed before any of the other steps are completed. The two types of grouping symbols you will learn to work with are parentheses and fraction bars.

ORDER OF OPERATIONS

1. Find the value of expressions within grouping symbols, such as parentheses and fraction bars.
2. Find the value of all powers.
3. Multiply and divide from left to right.
4. Add and subtract from left to right.

EXAMPLE 1

Find the value of each expression.

a. $5 \times (3 + 4) - 8$ **b. $30 + (1 + 3)^2 \div 2$**

SOLUTIONS

a. Find the value inside the parentheses. $5 \times (3 + 4) - 8 = 5 \times 7 - 8$
    Multiply. $5 \times 7 - 8 = 35 - 8$
    Subtract. $35 - 8 = 27$

b. Find the value inside the parentheses. $30 + (1 + 3)^2 \div 2 = 30 + 4^2 \div 2$
    Find the value of the power. $30 + 4^2 \div 2 = 30 + 16 \div 2$
    Divide. $30 + 16 \div 2 = 30 + 8$
    Add. $30 + 8 = 38$

EXAMPLE 2

Find the value of each expression.

a. $\dfrac{8-2}{1+1}$ **b.** $\dfrac{6^2}{3+6} - 2$

SOLUTIONS

a. Find the value of the numerator and the denominator. $\dfrac{8-2}{1+1} = \dfrac{6}{2}$

Divide the numerator by the denominator. $6 \div 2 = 3$

$\dfrac{8-2}{1+1} = 3$

> Remember that the fraction bar is the same as a division sign.

b. Find the value of the numerator and the denominator. $\dfrac{6^2}{3+6} - 2 = \dfrac{36}{9} - 2$

Divide. $\dfrac{36}{9} - 2 = 4 - 2$

Subtract. $4 - 2 = 2$

$\dfrac{6^2}{3+6} - 2 = 2$

EXPLORE!   **FACT PUZZLE**

Use the order of operations to find answers to these facts.

Step 1: The highest temperature (F°) ever recorded in the United States was in Death Valley, CA in 1913. What was the record temperature? $(6+4)^2 + 34$

Step 2: The Missouri River is the longest river in the United States. What is the length of the river in total miles? $2 \times 1000 + 8 \times 100 - 2 \times 130$

Step 3: As of 2010, how many active volcanoes are there in Hawaii? $\dfrac{26+24}{3+7} - 3$

Step 4: How many states were part of the United States in the year 1800? $3 \times (7-4) + 7$

Step 5: Florida has how many electoral votes for president? $(5+1)^2 - \dfrac{9^2}{4+5}$

EXERCISES

1. List the complete order of operations in words or pictures.

2. Give an example of each of the two grouping symbols used in this lesson.

3. Explain why it is necessary to have an order of operations for numerical expressions.

Find the value of each expression.

4. $\dfrac{8 + 12}{5 - 1}$

5. $7 \times (1 + 3)$

6. $(2 + 3)^2 - 12$

7. $21 \div 7 + 3 \times (6 + 2)$

8. $2 \times 20 \div 5 + 3$

9. $\dfrac{35 - 3}{2^3}$

10. $10^3 + 2 \times 100$

11. $5 \times (4 - 1)^2 - 10$

12. $3 \times 5 + \dfrac{3 + 7}{2}$

13. $\dfrac{3 + 4}{5 - 4} + \dfrac{8^2}{2}$

14. $9 \div 3 + 4 \times 2 - 11$

15. $3^3 + 2 \times (2 + 4)$

16. $8 + 2^3 \times 4 \times 3$

17. $\dfrac{5^2}{2} + 2.5$

18. $(10 - 3)^2 + 1^2$

Insert one set of parentheses in each numerical expression so that it equals the stated amount.

19. $2 \times 4 + 6 = 20$

20. $4 + 2 + 8 \div 2 = 9$

21. $15 + 1 \times 4 + 5 \times 2 = 74$

22. Abram makes $10 per hour at his job at the local gas station. He worked 4 hours on Monday, 5 hours on Tuesday and 3 hours on Saturday.
   a. How much did Abram earn?
   b. Explain two ways you could find the answer in **part a**.

23. Three friends order a pizza for $14, a bottle of soda for $3 and a large piece of cake to share for $4. The friends want to evenly split the costs. How much does each friend owe? Use mathematics to justify your answer.

REVIEW

Write each power in expanded form. Find its value.

24. 7^2

25. 3^3

26. 12^2

27. 2^4

28. 10^3

29. 1^5

Tic-Tac-Toe ~ Number Trick

Try this number trick:
> Pick a number.
> Add 5.
> Double the result.
> Subtract 4.
> Divide the result by 2.
> Subtract the number you started with.
> Your result is 3, right?

Try a different number in the trick above. Write out your computations in the order you do them. Do you get 3 again? Look at your written operations. Can you tell why this number trick works?

Josephine created the number trick below. Shown are two of her friends' computations. Copy the table and use words to describe each step of the number trick.

Description	Maria's Computations	August's Computations
Choose a number	7	1
	10	4
	20	8
	24	12
	12	6
	5	5

Try to create your own number trick with at least four steps. Show that your trick works by creating a table, like the one above, that shows at least three different starting numbers and a pattern for the solution on each.

Tic-Tac-Toe ~ Half-Life

The half-life of an element is the time it takes for half of the element to turn into another element. For example, Carbon-10 has a half-life of 20 seconds. If you began with 100 grams of Carbon-10, there would only be 50 grams of Carbon-10 left after 20 seconds. After 40 seconds, there would be 25 grams. Carbon-10 decays so fast that it cannot be found in nature. Choose one element listed below. Examine the amount of the element remaining after 10 half-lives. Make a chart, table or graph to illustrate how much of the original element is remaining at different times.

Element	Half-Life	Starting Amount
Promethium	17.7 years	1000 grams
Radon	3.8 days	200 grams
Francium	22 minutes	80 grams

NUMBER PROPERTIES

 Recognize and use the Commutative and Associative Properties.

There are some basic number properties in mathematics that you began following when you learned to add single-digit numbers. You will continue to use these properties all the way through calculus and beyond. These properties make it easier to do mental math by regrouping numbers or by moving numbers around in an expression.

EXPLORE!   **DISCOVERING PROPERTIES**

Step 1: Find the value of each expression.

      a. $2 + 5 + 7 + 3$ **b.** $7 + 3 + 2 + 5$ **c.** $5 + 3 + 2 + 7$

      What do you notice about the value of each expression?

Step 2: Find the value of each expression.

      a. $6 \times 2 \times 4$ **b.** $4 \times 2 \times 6$ **c.** $2 \times 6 \times 4$

      What do you notice about the value of each expression?

Step 3: Do you think you could rearrange the order of subtraction or division expressions and have the value of the expression always stay the same? Show at least one example to prove your answer.

Step 4: Follow the order of operations to find the value of each expression.

      a. $3 + (8 + 1)$ **b.** $(3 + 8) + 1$

      What do you notice about the value of each expression?

Step 5: Follow the order of operations to find the value of each expression.

      a. $(5 \times 7) \times 2$ **b.** $5 \times (7 \times 2)$

      What do you notice about the value of each expression?

Step 6: Follow the order of operations to find the value of each expression.

      a. $(60 \div 6) \div 2$ **b.** $60 \div (6 \div 2)$

      What do you notice about the value of each expression?

Step 7: Based on what you have discovered in this **Explore!**, for which operations $(+, -, \times, \div)$ can you move the numbers around in an expression, or group the numbers differently, and still get the same answer?

It is important to note that these properties only hold true for addition expressions containing only addition or multiplication expressions containing only multiplication. The Commutative Property and Associative Property do not work for subtraction or division.

Shown below are some ways to remember the names of these properties.

Commutative Property
Jake **COMMUTES** to work. He takes the bus to his job at the beginning of his day and when he is finished working, he heads home. He changes places each day, moving back and forth from home to work and back home again. The Commutative Property means to "change places in an expression".

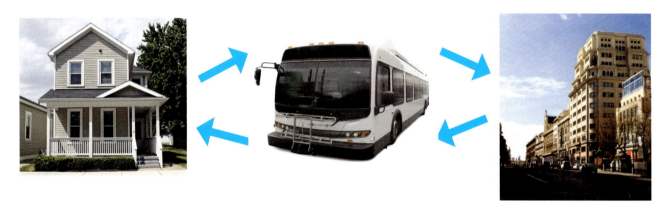

Associative Property
Kim, Natalie and Anna are friends at a local daycare. They **ASSOCIATE** with each other. One day, Kim and Natalie play together while Anna plays by herself. The next day, Natalie decides to play with Anna while Kim plays by herself. It is still the same group of friends, but they are grouped differently. They are associating with different friends on different days. The Associative Property means to "group numbers differently in an expression".

EXAMPLE 1

Determine if the expression in Column A is equal to the expression in column B. If the expressions are equal, identify the property shown.

COLUMN A	COLUMN B
a. $(3 + 2) + 9$	$3 + (2 + 9)$
b. $8 \div 4$	$4 \div 8$
c. $10 \times (4 - 2)$	$(10 \times 4) - 2$
d. 5×11	11×5

SOLUTIONS

a. Find the value of the expression in Column A. $(3 + 2) + 9 = 14$

Find the value of the expression in Column B. $3 + (2 + 9) = 14$

The expressions are equal. The numbers were grouped differently.
Property: Associative Property

b. Find the value of the expression in Column A. $8 \div 4 = 2$

Find the value of the expression in Column B. $4 \div 8 = 0.5$

The expressions are not equal.
The Commutative Property only works for addition and multiplication.

c. Find the value of the expression in Column A. $10 \times (4 - 2) = 20$

Find the value of the expression in Column B. $(10 \times 4) - 2 = 38$

The expressions are not equal.
The Associative Property only works for addition and multiplication. It also only works when you are using one operation at a time.

d. Find the value of the expression in Column A. $5 \times 11 = 55$

Find the value of the expression in Column B. $11 \times 5 = 55$

The expressions are equal. The numbers changed places.
Property: Commutative Property

EXERCISES

1. Explain the Commutative Property in your own words.

2. Write two expressions that are equal using the Commutative Property.

3. Explain the Associative Property in your own words.

4. Write two expressions that are equal using the Associative Property.

Identify the property shown by each mathematical statement.

5. $5 + (4 + 2) = (5 + 4) + 2$

6. $2 + 3 + 10 = 10 + 3 + 2$

7. $4 \times 9 = 9 \times 4$

8. $(9 \times 1) \times 7 = 9 \times (1 \times 7)$

Determine if the expression in Column A is equal to the expression in Column B. If the expressions are equal, identify the property shown.

COLUMN A	COLUMN B
9. $11 - (6 - 2)$	$(11 - 6) - 2$
10. $(4 \times 5) \times 3$	$4 \times (5 \times 3)$
11. $9 + 11$	$11 + 9$
12. $3 \div 1$	$1 \div 3$

13. Use the Commutative Property to write the numerical expression $4 \times 7 \times 3$ two other ways.

14. Use the Associative Property to write the numerical expression $8 + (2 + 5)$ another way.

15. The Commutative and Associative Properties allow you to move numbers around and group them so you can find the value of numerical expressions mentally. Use the expression $15 + 27 + 35$ to answer the following questions.
    a. Which two numbers in the expression would be easiest to add together first? Why?
    b. Rewrite the expression with the order you chose in **part a**.
    c. Find the value of the expression.

16. Maisha painted 27 square feet of a wall in the first hour. She painted 52 square feet in the second hour. She finished her painting project by painting 23 square feet in the last hour.
    a. How many total square feet did she paint?
    b. Use the two properties to show how you could group the numbers to make it easier to do mental math for this calculation.

Find the value of each expression.

17. $12 \times (4 - 2)$

18. $\dfrac{7 + 11}{2 + 4}$

19. $(1 + 9)^2 - 24$

20. $10 \times 3 \div 5 - 1$

21. $7 \times (2 - 1)^2 + 11$

22. $\dfrac{2 \times 24}{2^3}$

23. $2 \times 10 + 5 - 5^2$

24. $22 + 18 \div 3 \times 4$

25. $5 \times 8 + \dfrac{1 + 11}{2}$

TIC-TAC-TOE ~ PROPERTIES POSTER

The Associative and Commutative Properties are difficult for many students to remember. Create a poster that helps illustrate each property in a creative manner. Include an explanation of each property with words and numbers.

TIC-TAC-TOE ~ ORDER OF OPERATIONS POETRY

Poetry is an art form composed of carefully chosen words to express a greater depth of meaning. Poetry can be written about many different subjects, including mathematics. The order of operations is a key part of mathematics. Choose two or more types of poetry described below and write at least two poems about the order of operations.

Haiku

A Japanese poem composed of three unrhymed lines of five, seven, and five syllables.

Cinquain

Poetry with five lines.
Line 1 has one word (the title).
Line 2 has two words that describe the title.
Line 3 has three words that tell the action.
Line 4 has four words that express the feeling.
Line 5 has one word which recalls the title.

Acrostic

Poetry that tells about the word. It uses the letters of the word for the first letter of each line.

Vocabulary

Associative Property	cubed	order of operations
base	exponent	power
Commutative Property	grouping symbols	squared
	numerical expressions	

 Find values of expressions involving addition, subtraction, multiplication and division.
Write and compute expressions with powers.
Find values of expressions with exponents using the order of operations.
Find values of expressions with grouping symbols using the order of operations.
Recognize and use the Commutative and Associative Properties.

Lesson 1.1 ~ The Four Operations

Find the value of each expression.

1. $9 + 11 \times 2$

2. $39 \div 13 - 1 + 2$

3. $46 - 24 \div 8$

4. $2 \times 5 + 8 \times 1$

5. $4 \times 20 \div 10$

6. Three friends went to a movie. The movie tickets cost $12 each. The friends also purchased snacks to share for $17. What was the total cost for the evening at the movies? Use mathematics to justify your answer.

Lesson 1.2 ~ Powers and Exponents

Write the numerical expression as a power.

7. $4 \times 4 \times 4$

8. $7 \times 7 \times 7 \times 7 \times 7 \times 7$

9. 3×3

10. $2 \times 2 \times 2 \times 2 \times 2 \times 2 \times 2$

11. $9 \times 9 \times 9 \times 9 \times 9$

12. $15 \times 15 \times 15$

Write each power in expanded form and find the value.

13. 4^3

14. 1^4

15. 11^2

Determine which power has the greater value.

16. 4^2 or 3^3

17. 6^2 or 2^6

18. $\left(\frac{1}{2}\right)^2$ or $\left(\frac{1}{2}\right)^3$

19. The volume of a cube can be calculated using the formula *length × width × height*.
   a. Write the volume of the cube shown in expanded form.
   b. Write the volume of the cube as a power.
   c. Find the value of the volume of the cube.

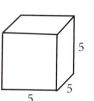

Lesson 1.3 ~ Order of Operations with Powers

Find the value of each expression.

20. $8 + 5^2 \times 4$

21. $36 \div 3 + 2^3 - 5$

22. $17 - 1^8 \times 6$

23. $8^2 \div 2 + 68$

24. $2 \times 6^3 - 6$

25. $3 + 2^3 \times 3^2 - 10$

26. In each numerical equation one number should be squared (have a power of two) so that it equals the stated amount. Rewrite the equation with the appropriate number squared. Use mathematics to show that your answer is correct.
   a. $2 \times 6 + 8 = 76$
   b. $9 + 10 \div 2 = 59$

Lesson 1.4 ~ Order of Operations with Grouping Symbols

Find the value of each expression.

27. $\dfrac{7 + 2}{4 - 1}$

28. $5 \times (6 + 3)$

29. $(3 + 7)^2 - 6$

30. $20 \div 4 + 3 \times (12 - 7)$

31. $50 \times 2 \div (5 + 5)$

32. $\dfrac{10 + 4^2}{2}$

33. Insert one set of parentheses in each numerical expression so it equals the stated amount. Rewrite each expression with the parentheses in the appropriate place. Use mathematics to show that your answer is correct.
   a. $3 \times 5 + 8 = 39$
   b. $5 + 3 \times 4 + 12 \div 4 = 35$

34. Tori makes $9 per hour at her job at the local deli. This week she worked 2 hours on Tuesday. On Thursday she worked 4 hours. On Saturday she worked 7 hours.
   a. How much did Tori earn this week?
   b. Explain two ways you could find the answer in **part a**.

Lesson 1.5 ~ Number Properties

Identify the property shown by each mathematical statement.

35. $5 + 14 = 14 + 5$

36. $2 \times (3 \times 6) = (2 \times 3) \times 6$

37. $(7 + 8) + 9 = 7 + (8 + 9)$

38. $8 \times 3 \times 12 = 12 \times 8 \times 3$

Determine if the expression in Column A is equal to the expression in Column B. If the expressions are equal, identify the property shown.

	COLUMN A	COLUMN B
39.	$(9 + 2) + 7$	$9 + (2 + 7)$
40.	4×12	12×4
41.	$8 - (4 - 1)$	$(8 - 4) - 1$

42. Use the Commutative Property to write the numerical expression $5 + 7 + 8$ two other ways.

43. Use the Associative Property to write the numerical expression $13 + (6 + 3)$ another way.

TIC-TAC-TOE ~ DISTANCES TO PLANETS

Scientific notation is a method used by scientists and mathematicians to express very large numbers. Scientific notation is an exponential expression using a power of 10.

$$N \times 10^P$$

Use the following process to convert a large number into scientific notation:

Step 1: Locate the decimal point and move it left so there is only one non-zero digit to its left. This number represents the value of N.

Step 2: Count the number of places that you moved the decimal point in **Step 1**. This number represents the value of P.

Example: Convert 43,000 to scientific notation:

1. Move the decimal point to the left so there is only one non-zero digit to the left of the decimal point.

$$43{,}000 \rightarrow 4.3$$

2. Count how many places the decimal point moved to determine the power of 10.

43000   The decimal point has moved four places.

43,000 in scientific notation is 4.3×10^4

There are eight planets in the solar system, including Earth. Use resources to determine the distance (in miles) of each planet from Earth. Write the distance in regular notation (called standard notation) and in scientific notation.

CAREER FOCUS

TRACEY
CIVIL ENGINEER

I am a civil engineer. I design and supervise the building of bridges and buildings. I work with architects and other types of engineers to provide a complete bridge or building. Some civil engineers work in research or teach other engineers. Most civil engineers work in large industrial cities, but some projects may be in isolated places or foreign countries. Although I live and work in a large city, I often design and help supervise buildings all over the state or in other parts of the country. A few of my projects include designing an apartment complex, designing and supervising the building of a new library at a university and designing and supervising the building of a new airplane hangar.

As a civil engineer, I use math everyday. I work with fractions when I am looking at blueprints of my projects. I use formulas to help me determine if the building will be "sound", which means making sure it will stay standing for a long time. I have to provide mathematical evidence to support my designs on all my projects. I often use high-level mathematics such as vectors and quadratic equations.

In order to become a civil engineer, I received a Bachelor's of Science in Engineering. This took me five years of college but some people can finish the program in four years. Most civil engineers start their career making about $44,000 annually. The middle half of all civil engineers earn between $51,000 and $80,000 annually.

I love my job because I get to watch a project go from start to finish. I know that the buildings I design will stay standing for a long time because I have checked and double-checked all of my math to verify that the building plans are sound. I can drive around and see buildings that I helped design.

CORE FOCUS ON INTRODUCTORY ALGEBRA
BLOCK 2 ~ ALGEBRAIC EXPRESSIONS

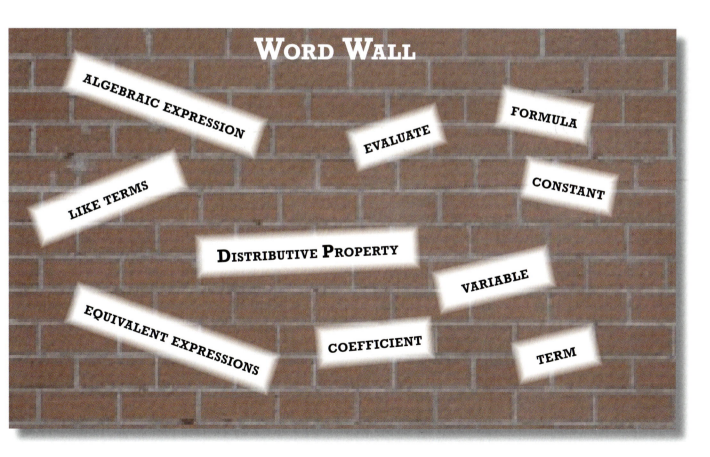

WORD WALL

ALGEBRAIC EXPRESSION

EVALUATE

FORMULA

LIKE TERMS

CONSTANT

DISTRIBUTIVE PROPERTY

VARIABLE

EQUIVALENT EXPRESSIONS

COEFFICIENT

TERM

BLOCK 2 ~ ALGEBRAIC EXPRESSIONS
TIC-TAC-TOE

POPULATION DENSITY

Determine the population density of different areas of the world.

See page 52 for details.

DREAM SCHOOL

Create a floor plan for a one-story school. Find the area of each room and the total square footage of the school.

See page 47 for details.

EXPRESSIONS GAME

Create a matching game by creating game cards with equivalent expressions.

See page 61 for details.

BANK INTEREST

Find the interest rate for a savings account at local banks. Use the compound interest formula to determine the amount of interest you could earn.

See page 48 for details.

WRITING EXPRESSIONS

Write complex algebraic expressions which include powers and grouping symbols. Evaluate the expressions.

See page 37 for details.

GROCERY SHOPPING

Take a shopping list to the grocery store and use the Distributive Property to determine the total cost for the items on the list.

See page 57 for details.

CHILDREN'S BOOK

Write a children's story about variables. Use one formula from Lesson 2.4 in your story.

See page 48 for details.

OPERATIONS VOCABULARY

Make a poster of vocabulary that is used for each of the four operations (+, −, ×, ÷).

See page 32 for details.

GEOMETRY FORMULAS

Research formulas for geometric shapes such as a rhombus, trapezoid and parallelograms. Determine the areas of given shapes.

See page 42 for details.

VARIABLES AND EXPRESSIONS

 Write expressions involving variables.

Jenny shared cookies with her friends. Each cookie contained 55 calories. The chart below shows the number of cookies each friend ate and the calories each friend consumed.

Friend's Name	Number of Cookies Eaten	Calculation	Calories Consumed
Sam	4	4×55	220
Jake	7	7×55	385
Mia	2	2×55	110
Julie	9	9×55	495
Chris	c	$c \times 55$	$c \times 55$

Chris did not remember how many cookies he ate. A **variable** is a letter that stands for a number. In this case, the variable stands for the number of cookies Chris consumed. An **algebraic expression** is a mathematical expression that contains numbers, operations (such as add, subtract, multiply or divide) and variables. The algebraic expression that represents the number of cookies Chris ate is $c \times 55$.

In order to write algebraic expressions you must be able to translate mathematical words into symbols.

EXPLORE!   TRANSLATE THOSE WORDS

Step 1: Fold a piece of paper in half vertically. Open up your paper and then fold your paper in half horizontally. Open the paper up and lay it flat on your desk.

Step 2: Label each section of your paper as shown.

Step 3: Place each word or phrase from the list below into one of the sections on your paper. Be prepared to explain your placement of each word or phrase.

Addition	Subtraction
Multiplication	**Division**

less than   sum   decreased by

quotient   more than   divided by

times   difference   increased by

minus   product   fewer than

Step 4: Draw the mathematical symbol used to represent each operation at the bottom corner of each box.

Step 5: Eight algebraic expressions are listed below. There are two representing each operation. Place the algebraic expressions into the box that describes each operation.

$x - 4$ $\qquad\qquad$ $6 \div p$ $\qquad\qquad$ $5 \times w$ $\qquad\qquad$ $14 + m$

$30 - g$ $\qquad\qquad$ $10k$ $\qquad\qquad$ $y + 11$ $\qquad\qquad$ $\dfrac{h}{2}$

Step 6: In each box, write your own algebraic expression that uses the operation listed at the top of that box. Compare your expressions with a classmate. Are the expressions exactly the same? Should they be?

Variables are an important part of mathematics. One of the most commonly used variables is the letter x. Because the letter x looks so much like the multiplication sign ×, you will not use × to represent multiplication when working with variables. For the rest of this book and in future math classes, you will show multiplication by using one of the methods below. Each one shows $4 \times y$.

> **METHODS FOR SHOWING MULTIPLICATION**
>
> Dot: $4 \cdot y$
> Parentheses: $4(y)$
> Number and adjacent variable: $4y$

EXAMPLE 1

Write an algebraic expression for each phrase.
a. five times k
b. seven more than x
c. f divided by two
d. a number y decreased by seventeen
e. eleven plus the product of four and w

SOLUTIONS

a. $5k$ $\quad$ or $\quad$ $5(k)$ $\quad$ or $\quad$ $5 \cdot k$

> When multiplying a number and variable, write the number first.

b. $x + 7$

c. $f \div 2$ $\quad$ or $\quad$ $\dfrac{f}{2}$

d. $y - 17$

e. $11 + 4w$ $\quad$ or $\quad$ $11 + 4(w)$ $\quad$ or $\quad$ $11 + 4 \cdot w$

EXAMPLE 2

Write a phrase for each algebraic expression.
a. $12 + t$ b. $8x$ c. $u - 5$ d. $3x - 2$

SOLUTIONS

a. the sum of twelve and t
b. the product of eight and x
c. five less than u
d. two less than three times x

> Answers may vary. Can you think of other answers for each part of this example?

EXERCISES

1. Michael's allowance is always $3 less than his older brother Keith's allowance. Copy the table and fill in the missing boxes.

Keith's Allowance	Calculation	Michael's Allowance
$14	14 − 3	$11
$20	20 − 3	
	25 − 3	
$31		
x	$\longrightarrow$	

2. Which of the following is not an appropriate way of writing "six times x" as an algebraic expression? Why?

$6x$ $6 \times x$ $6 \cdot x$ $6(x)$

Write an algebraic expression for each phrase.

3. the sum of y and eleven

4. nine times m

5. the product of z and seven

6. the quotient of c divided by three

7. five more than x

8. four subtracted from p

9. thirty-one less than w

10. fourteen divided by b

11. a number n decreased by thirteen

12. five times a number x

Write a phrase for each algebraic expression.

13. $y - 2$ **14.** $3 \cdot d$ **15.** $60 - x$

16. $p + 16$ **17.** $\frac{r}{5}$ **18.** $2w + 1$

19. Write three different phrases for $x + 5$.   **20.** Write three different phrases for $10 - p$.

21. A horse's heart rate is about 38 beats per minute.
  a. How many times will a horse's heart beat in 3 minutes?
  b. How many times will a horse's heart beat in 10 minutes?
  c. How many times will a horse's heart beat in b minutes?

22. Yoshi gets half as many tardy slips as Ryan gets.
  a. If Ryan got 12 tardy slips in March, how many did Yoshi get?
  b. If Ryan got 26 tardy slips in the first semester, how many did Yoshi get?
  c. Ryan got t tardy slips all year. Write an algebraic expression that shows how many tardy slips Yoshi received.

23. Francine's Fruit Stand sells apples and pears by the pound. The pears cost $1.25 per pound more than the apples.
  a. The apples cost $0.75 per pound. How much do the pears cost per pound?
  b. The apples cost x dollars per pound. What algebraic expression represents the cost of a pound of pears?
  c. The pears cost y dollars per pound. What algebraic expression represents the cost of a pound of apples?

REVIEW

Write each power in expanded form and find the value.

24. 11^2

25. 1^6

26. 4^3

27. 20^2

28. 2^4

29. 0^5

Determine if the expression in Column A is equal to the expression in Column B. If the expressions are equal, identify the property shown.

COLUMN A	COLUMN B
30. 4×15	15×4
31. $80 \div (8 \div 4)$	$(80 \div 8) \div 4$
32. $(3 + 7) + 10$	$3 + (7 + 10)$

TIC-TAC-TOE ~ OPERATIONS VOCABULARY

There are many different vocabulary words which can be used for the four basic mathematical operations ($+$, $-$, $\times$, $\div$). Design a poster that lists as many words as you can find for each operation. Research each operation to find more vocabulary than just the words given in this book. Include both mathematical expressions and word phrases on your poster as examples of each operation.

EVALUATING EXPRESSIONS

Evaluate algebraic expressions.

Gary delivers newspapers every morning. Each day he delivers papers to his customers he earns $14. The algebraic expression that represents his total earnings based on the number of days (*d*) he has delivered papers is 14*d*.

You can **evaluate** an algebraic expression by substituting a number for the variable to find its value. Some algebraic expressions will have more than one variable. If this is the case, you will replace each variable with a given number. After replacing the variable, or variables, with numbers, you will use the order of operations to find the value of the expression. The value of an algebraic expression changes depending on the value of the variable.

EVALUATING AN EXPRESSION
1. Rewrite the expression by replacing the variable(s) with the given value(s).
2. Follow the order of operations to find the value of the expression.

Gary delivered newspapers for eight days. Substitute the number 8 in the algebraic expression for the letter *d*.
$$14d \rightarrow 14(8) = \$112$$

The value of the algebraic expression, $112, represents how much money Gary has made after delivering the paper for eight days.

EXAMPLE 1

Evaluate each algebraic expression.

a. $p + 4$ when $p = 7$

b. $9x$ when $x = 3$

c. $\dfrac{y}{3} + 1$ when $y = 24$

SOLUTIONS

a. Rewrite the expression with 7 in the place of *p*.
Compute the value of the expression.

$p + 4 \rightarrow 7 + 4$
$7 + 4 = 11$

b. Rewrite the expression by substituting 3 for *x*.
Compute the value of the expression.

$9x \rightarrow 9(3)$
$9(3) = 27$

c. Substitute 24 for *y*.

$\dfrac{y}{3} + 1 \rightarrow \dfrac{24}{3} + 1$

Use the order of operations to evaluate.

$\dfrac{24}{3} + 1 = 8 + 1 = 9$

It is possible for algebraic expressions to include multiple variables. Follow the same process by substituting the value of each variable in the correct places and then find the value of the expression.

EXAMPLE 2

Evaluate each algebraic expression when $x = 3$, $y = 10$ and $z = 5$.

a. $4x - y$

b. $\dfrac{y}{2z} + x$

SOLUTIONS

a. Substitute 3 for x and 10 for y.
Evaluate using the order of operations.

$4x - y \rightarrow 4(3) - 10$
$4(3) - 10 = 12 - 10 = 2$

b. Substitute 10 for y, 5 for z and 3 for x.

$\dfrac{y}{2z} + x \rightarrow \dfrac{10}{2(5)} + 3$

Use the order of operations to evaluate.

$\dfrac{10}{2(5)} + 3 = \dfrac{10}{10} + 3 = 1 + 3 = 4$

Different values can be substituted for the variable in an expression. A table is used to organize mathematical computations. The substituted values are called the input values.

INPUT VALUES   **ALGEBRAIC EXPRESSION**

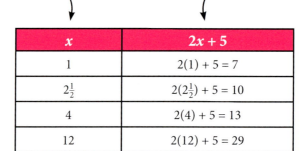

x	$2x + 5$
1	$2(1) + 5 = 7$
$2\frac{1}{2}$	$2(2\frac{1}{2}) + 5 = 10$
4	$2(4) + 5 = 13$
12	$2(12) + 5 = 29$

EXAMPLE 3

Shari's Taxi Service charges customers an initial fee of $5 plus $0.50 per mile driven for each ride. The algebraic expression representing the total cost is $5 + 0.50m$, where m is the total number of miles driven during one ride. Use a table to show how much rides of 6 miles, 15 miles and 24 miles would cost Shari's customers.

SOLUTION

Miles driven (m)	Cost ($5 + 0.50m$)
6	$5 + 0.50(6) = 5 + 3 = \$8.00$
15	$5 + 0.50(15) = 5 + 7.50 = \$12.50$
24	$5 + 0.50(24) = 5 + 12 = \$17.00$

EXERCISES

Evaluate each expression.

1. $x - 7$ when $x = 18$

2. $12b$ when $b = 4$

3. $4y - 1$ when $y = 5$

4. $50 - 3k$ when $k = 10$

5. $\frac{1}{2} + f$ when $f = 18$

6. $36 \div w$ when $w = 9$

7. $5d + 12$ when $d = 2$

8. $0.5h + 2$ when $h = 8$

9. $\frac{v}{5} + 7$ when $v = 40$

10. Julius starts with \$40 in his savings account. Each week he adds \$6 to his account. The algebraic expression that represents the total money Julius has in his account is $40 + 6w$ where w represents the number of weeks he has been adding money to his account.

    a. How much money will Julius have in his account after 3 weeks?

    b. How much money will Julius have in his account after 12 weeks?

    c. After a year, how much money should Julius have in his savings account? Show all work necessary to justify your answer.

11. Most teenagers take approximately 21 breaths per minute.

    a. How many breaths does a typical teenager take in 20 minutes?

    b. How many breaths does a typical teenager take in an hour?

    c. How many breaths does a typical teenager take in m minutes?

12. Johanna is going to the State Fair this summer. The fair charges a \$9 admission fee plus \$4 per ride.

    a. Write an algebraic expression that represents Johanna's total cost to attend the State Fair when she goes on r rides.

    b. Evaluate the total cost for Johanna to go on 2, 5 or 11 rides. Display this information in a table.

Evaluate each expression when $a = 6$, $b = 2$ and $c = 30$. Show all work necessary to justify your answer.

13. $2b + c$

14. $3(a + b)$

15. $\frac{c}{a} + b$

16. $\frac{12 + a}{b}$

17. $a - 2b + 3c$

18. $5ab$

19. Write an expression that has a value of 54 that uses the variables a, b and c as defined above. Use words and/or numbers to show how you determined your expression.

20. Create three different expressions using x, y and/or z that have a value of 100 if $x = 5$, $y = 2$ and $z = 25$. Use mathematics to prove your expressions have a value of 100.

Copy each table. Complete each table by evaluating the given expression for the values listed.

21.

x	$6x + 3$
0	
$\frac{1}{2}$	
3	
8	
20	

22.

x	$\dfrac{5x - 3}{2}$
1	
5	
6	
9	

23. James is planning an ice cream social. He needs to purchase containers of ice cream and packages of cones. The ice cream costs \$3.50 per container and the packages of cones cost \$2.00 each. James uses the algebraic expression $3.50x + 2.00y$ to calculate his total expenses.
   a. What does the x variable stand for? The y variable?
   b. James ends up buying 4 containers of ice cream and 3 packages of cones. How much will he spend? Use mathematics to justify your answer.

24. A rock was thrown from a cliff. The height of the rock from the base of the cliff (in feet) can be found using the expression $80 + 64t - 16t^2$ where t is the number of seconds since the rock was thrown.
   a. How high was the rock after 3 seconds?
   b. How high off the ground was the rock before it was thrown? Use words and/or numbers to show how you determined your answer.

25. Toby calculated the value of the expression $5y + 7$ when $y = 4$. He said the answer was 61. He made a common mistake. Write the expression by substituting 4 for y. What mistake did Toby make? Give the correct value of the expression.

REVIEW

Insert one set of parentheses in each numerical expression so that it equals the stated amount. Rewrite the problem with the parentheses in the appropriate places. Use mathematics to show that your answer is correct.

26. $2 \cdot 6 + 3 = 18$

27. $4 \cdot 5 + 20 \div 10 = 10$

28. $1 + 30 \div 5 \cdot 2 - 1 = 7$

Find the value of each expression.

29. $\dfrac{13 + 5}{(4 - 3)^2}$

30. $4 \cdot 6 + (12 - 3)^2$

31. $(2 + 5)^2 - (1 + 4)^2$

32. Pete's work to find the value of $2 + 6 \times (7 - 1)$ is shown at the right. Explain whether or not his work is correct. If it is incorrect, show Pete how to find the correct value.

Problem: $2 + 6 \times (7 - 1)$

Step 1: $2 + 6 \times 6$

Step 2: 8×6

Answer: 48

TIC-TAC-TOE ~ WRITING EXPRESSIONS

Word phrases for algebraic expressions get more difficult as expressions get more complex. The expressions in this Block only use the four basic operations: add, subtract, multiply and divide. As you have learned, mathematical expressions may also contain powers and grouping symbols such as parentheses and fraction bars.

When grouping symbols are used, the word phrase for the expression should contain "the quantity of". When exponents are used in an expression, the word phrase may contain words such as "squared", "cubed" or "to the power of…"

Examples:

Expression	Possible Phrase
$4(x + 3)$	four times the quantity of x plus 3
$x^3 - 9$	nine less than x cubed
$(x + 7)^2$	the quantity of x plus seven squared
$\dfrac{x - 1}{8}$	the quantity of x minus 1 divided by eight

Write a word phrase for each algebraic expression.

1. $(x + 5)^2$

2. $8(x - 11)$

3. $4 + x^4$

4. $\dfrac{10}{x + 1}$

5. $2(x + 5) + x^3$

6. $(x - 6)^2 + 5x$

Write an algebraic expression for each word phrase.

7. the quantity of x minus four squared

8. nine times the quantity of x minus six

9. seventy-two divided by the quantity of x plus one

10. the quantity of seventy divided by x decreased by the quantity of x minus six cubed

11. Evaluate #7–10 when $x = 7$. If completed correctly, all answers should be the same.

EVALUATING GEOMETRIC FORMULAS

 Use geometric formulas to find area, perimeter and volume.

A formula is an algebraic equation that shows the relationship among specific quantities. Geometric formulas are used to evaluate quantities such as perimeter, area or volume. In past grades, you learned that the area of a rectangle is equal to the length of the rectangle times the width. This can be written as the geometric formula Area = *lw*. In this lesson you will examine and evaluate a variety of different geometric formulas.

EXPLORE!   **USING FORMULAS**

Area = $\frac{1}{2}bh$

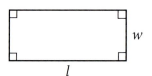

Area = *lw*
Perimeter = 2*l* + 2*w*

Volume = *lwh*
Surface Area = 2(*lw* + *wh* + *hl*)

Step 1: Look at the three shapes above. Can you name the shapes? Identify what you think the variables in each of the geometric formulas represent. Compare with your classmates.

Step 2: On the rectangle above, let *l* = 6 and *w* = 2. Use the given geometric formula to find the area of this rectangle.

Step 3: Using the values of *l* and *w* from **Step 2**, find the perimeter of the rectangle using the perimeter formula.

Step 4: Find the area of the triangle when *b* = 10 and *h* = 3.

Step 5: Find the volume of the rectangular prism above when *l* = 5, *w* = 4 and *h* = 3.

Step 6: Find the surface area of the rectangular prism using the values for *l*, *w* and *h* in **Step 5**.

Many people calculate area and perimeter in their jobs. A carpet layer must know the area of the floor he is covering in order to bring the right amount of carpet. A fence-builder must know how to calculate perimeter in order to charge the right price for building a fence. When someone wants to paint a house, he or she must know the surface area in order to know how much paint to buy. Can you think of other careers that would require using geometric formulas?

EXAMPLE 1

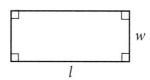

Area = $\frac{1}{2}bh$ Area = lw Area = πr^2
                          Perimeter = $2l + 2w$ Circumference = $2\pi r$

Evaluate each formula when given the values for the variables.
a. Find the area of a triangle when $b = 12$ and $h = 4$.
b. Find the area and perimeter of a rectangle when $l = 9$ cm and $w = 5$ cm.
c. Find the area and circumference of a circle when $r = 3$ inches. Use 3.14 for π.

SOLUTIONS   **a.** Use the formula for the area of a triangle.   Area = $\frac{1}{2}bh$
$$= \frac{1}{2}(12)(4)$$
$$= \frac{1}{2}(48)$$

> Area is labeled with square units. The units on perimeter are not squared.

$$= 24 \text{ square units}$$

b. Use the formula for the area of a rectangle.   Area = lw
$$= (9)(5)$$
$$= 45 \text{ square centimeters}$$

Use the formula for the perimeter of a rectangle. Perimeter = $2l + 2w$
$$= 2(9) + 2(5)$$
$$= 18 + 10$$
$$= 28 \text{ centimeters}$$

c. Use the formula for the area of a circle.   Area = πr^2
$$= (3.14)(3^2)$$
$$= (3.14)(9)$$
$$= 28.26 \text{ square inches}$$

Use the formula for the circumference of a circle. Circumference = $2\pi r$
$$= 2(3.14)(3)$$
$$= 2(9.42)$$
$$= 18.84 \text{ inches}$$

EXAMPLE 2   **A cereal manufacturer is introducing its new breakfast cereal, Strawberry Crunchettes. The box the cereal will be sold in has a length of 12 inches, a width of 4 inches and a height of 15 inches.**

a. Find the volume of the box using the volume formula:
   $V = lwh$. This represents how much cereal the box can hold in cubic inches.

b. Find the surface area of the box using the surface area formula: Surface Area = $2(lw + wh + hl)$. This represents the amount of cardboard needed to make the cereal box in square inches.

EXAMPLE 2
SOLUTIONS
(CONTINUED)

a. The volume of the box is found by substituting 12 for l, 4 for w and 15 for h.

$V = lwh$
$= (12)(4)(15)$
$= 720$ cubic inches

The volume of the cereal box is 720 cubic inches.

b. The surface area of the box is found by substituting the same numbers into the surface area formula.

Surface Area $= 2(lw + wh + hl)$
$= 2(12 \cdot 4 + 4 \cdot 15 + 15 \cdot 12)$
$= 2(48 + 60 + 180)$
$= 2(288)$
$= 576$ square inches

The surface area of the cereal box is 576 square inches.

EXERCISES

1. The area of a rectangle is found using the geometric formula Area $= lw$. The perimeter of a rectangle is found using the geometric formula $P = 2l + 2w$.
 a. Find the area and perimeter of a rectangle when $l = 5$ and $w = 3$.
 b. Find the area and perimeter of a rectangle when $l = 2$ and $w = 0.5$.
 c. Find the area and perimeter of a rectangle when $l = 24$ and $w = 10$.

2. The front and back walls of an A-Frame cabin are triangular in shape. A painter wants to determine the area of the triangular front of the cabin in order to know how much paint to buy. The base of the triangle (b) is 16 feet. The height of the triangle (h) is 20 feet. Find the area of the front of the A-Frame cabin. (Area of a triangle $= \frac{1}{2}bh$)

Evaluate each area, perimeter or circumference using the geometric formulas below.

Area $= \frac{1}{2}bh$

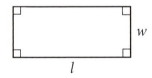

Area $= lw$
Perimeter $= 2l + 2w$

Area $= \pi r^2$
Circumference $= 2\pi r$

3. area of a triangle when $b = 6\ m$ and $h = 7\ m$

4. area of a circle when $r = 5\ cm$ (Use 3.14 for π.)

5. perimeter of a rectangle when $l = 11$ and $w = 3$

6. area of a triangle when $b = 50\ ft$ and $h = 10\ ft$

7. area of a rectangle when $l = 11$ and $w = 3$

8. perimeter of a rectangle when $l = 32$ and $w = 14$

9. circumference of a circle when $r = 10$ inches (Use 3.14 for π.)

Evaluate the surface area or volume of a rectangular prism using the geometric formulas below.

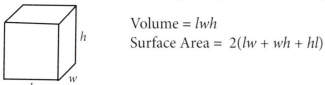

Volume = *lwh*
Surface Area = 2(*lw* + *wh* + *hl*)

10. volume of a rectangular prism when *l* = 4 *in*, *w* = 3 *in* and *h* = 5 *in*

11. volume of a rectangular prism when *l* = 6, *w* = 6 and *h* = 6

12. surface area of a rectangular prism when *l* = 2, *w* = $\frac{1}{2}$ and *h* = 2

13. surface area of a rectangular prism when *l* = 3 *cm*, *w* = 2 *cm* and *h* = 7 *cm*

14. MegaMakers Candy Company is creating a new box for their top-selling candy, Choco Chunks. The box the company has created has a length (*l*) of 4 inches, a width (*w*) of 2 inches and a height (*h*) of 5 inches.
    a. Find the volume of the box (how much space is inside the box) in cubic inches.
    b. Find the surface area of the box (how much cardboard is needed to make the box) in square inches.
    c. Each piece of the Choco Chunks candy is 0.5 cubic inches. How many pieces of Choco Chunks will fit in the new box?

15. Kyle and his friends race around a circular track. The radius of the track is 40 meters. Kyle wants the race to be 1,500 meters long. Approximately how many times around the track is this? Use 3.14 for π. Show all work necessary to justify your answer.

16. The volume of a sphere is found using the formula V = $\frac{4}{3}\pi r^3$. A beach ball has a radius (*r*) of 10 inches. What is the volume of the ball? Use 3.14 for π.

17. Ahn drew a figure made of four congruent triangles and a square. Find the total area of the figure. Use words and/or numbers to show how you determined your answer.

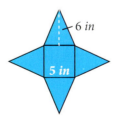

6 *in*

5 *in*

18. Ivan found the perimeter of a rectangle with a width of 10 inches and a length of 4 inches. His work is shown below. Explain how you know his answer is correct or incorrect.

> Perimeter = 2l + 2w
>   = 2(4) + 2(10)
>   = 6 + 12
>   = 18 inches

19. A rectangle has a width that is half as long as its length. The length of the rectangle is $3\frac{1}{2}$ feet. What is the area of the rectangle? Use mathematics to justify your answer.

REVIEW

Evaluate each expression when $x = 1$, $y = 15$ and $z = 5$.

20. $x + y + z$

21. $2(y - x)$

22. $\frac{y}{z}$

23. $14 + 2y - z$

24. $5(x + z) + y$

25. $3z + 2y$

TIC-TAC-TOE ~ GEOMETRY FORMULAS

Mathematicians have developed formulas to find the areas of many different geometric shapes. Quadrilaterals are geometric shapes which have four sides. Three types of quadrilaterals are trapezoids, rhombuses and parallelograms.

Step 1: Find the definition for a trapezoid, rhombus and parallelogram. Draw an example.

Step 2: Find the formulas used to calculate the area of a trapezoid, rhombus and parallelogram.

Step 3: Copy each shape below onto paper. Write the name for each shape. Find the area of each shape.

1.
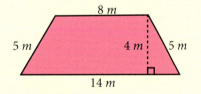

2.
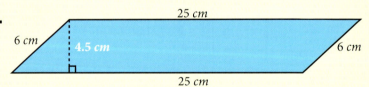

3.
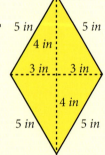

4.
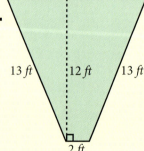

Step 4: Draw a new rhombus, parallelogram and trapezoid. Find the length of each side, height or diagonal to the nearest tenth of a centimeter. Find the area of each of your quadrilaterals.

EVALUATING MORE FORMULAS

Use formulas in a variety of situations.

Formulas are equations that show relationships between variables. In **Lesson 2.3** you saw how formulas are used in geometry. In this lesson you will see how formulas are used in all types of situations in real life. For example, formulas can be used to determine distance traveled, calculate interest earned in a savings account and calculate batting averages.

EXPLORE!   EARNING INTEREST

Money deposited into a bank account is called the principal. The bank pays interest on the money for as long as it is in the bank. Simple interest is calculated based on the principal. The formula used to calculate the amount of simple interest earned is $I = prt$. The interest is represented by I, p is the principal, r is the interest rate per year as a decimal and t is the time in years.

Step 1: Nicholas deposited $400 in an account and left it there for 3 years. His account earns 5% per year. Match each number to a variable in the simple interest formula:

$$p = \qquad r = \qquad t =$$

Step 2: The rate must be converted from a percent to a decimal before it is put into the formula. To convert a percent to a decimal, divide the percent by 100. What is 5% as a decimal?

Step 3: Substitute the values for p, t and the decimal value of r into the equation $I = prt$. Evaluate the equation. How much interest did Nicholas earn?

Step 4: Tracey found a savings account that earns 6% per year. She deposited $200 in an account and left the money there for 5 years. How much did she earn? How much total money will she have after 5 years?

Step 5: Ian was given $500. He invested it in a savings account for 6 months at 8% per year. To calculate his interest, he substituted the values into the equation. Look at his calculation below. Did he calculate correctly? If not, how much money did he really earn?

$$I = prt = (500)(0.08)(6) = \$240$$

Another formula used in math and science is $d = rt$. The variable d represents distance, r is rate (or speed) and t is time. This formula can be used to determine the distance a car travels over a certain amount of time or the distance a jogger runs in a specific amount of time.

EXAMPLE 1

A family traveled from Midland, Texas to Ranger, Texas in a small sedan. They traveled for 4 hours at a speed of 58 miles miles per hour. How many miles did they drive from Midland to Ranger?

SOLUTION

Identify which number to substitute for each variable in the formula.

distance = d = ?
rate (speed) = r = 58
time = t = 4

Substitute the values into the formula $d = rt$.

$d = rt$
$d = (58)(4)$
$d = 232$ miles

The family traveled 232 miles.

EXAMPLE 2

Petrik jogs at a speed of 7 miles per hour. He jogs for 30 minutes. How far has he traveled?

SOLUTION

Petrik's speed is in miles per hour. The amount of time is in minutes. Convert the minutes to hours before substituting values into the formula.

60 minutes = 1 hour
so
30 minutes = $\frac{1}{2}$ hour

Identify which number to substitute for each variable in the formula.

distance = d = ?
rate (speed) = r = 7
time = $t = \frac{1}{2}$

Substitute the values into the formula $d = rt$.

$d = rt$
$d = (7)\left(\frac{1}{2}\right)$
$d = 3\frac{1}{2}$

Petrik jogged $3\frac{1}{2}$ miles.

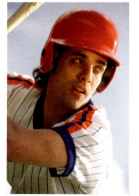

In baseball, each batter has a batting average. The batting average is defined as the ratio of hits to 'at bats'. The formula used to calculate the batting average is: $B = \frac{h}{a}$, where B is the batting average, h is the number of hits and a is the number of 'at bats'.

EXAMPLE 3

A professional baseball player had 555 'at bats' in one season. He had a total of 189 hits during that season. What was his batting average?

SOLUTION

Identify which number to substitute for each variable in the formula.

batting average = B = ?
hits = h = 189
'at bats' = a = 555

Substitute the values into the formula $B = \frac{h}{a}$.

$B = \frac{h}{a}$

$B = \frac{189}{555} \approx 0.341$

The baseball player had a batting average of 0.341.

> Round batting averages to the nearest thousandth.

EXERCISES

Use the simple interest formula, $I = prt$, to evaluate the amount of interest earned.

1. Find the amount of interest when p = $100, r = 0.04 and t = 6 years.

2. Find the amount of interest when p = $2000, r = 0.07 and t = 2 years.

3. Vinny deposited $1,400 in an account for 3 years at 9% interest. How much money did he earn?

4. Jackie deposited $500 in an account for 10 years at 6% interest.
   a. How much money did she earn?
   b. She added the interest she earned to the money in her savings account. She left the money in the account for one more year at 6%. How much additional money will she earn in the last year?

5. David left $300 in an account that earns 5% interest. When he collected his interest, he had earned $60 in interest. Did he leave his money in the account for 2, 4 or 6 years? Explain how you know your answer is correct.

6. Loans are CHARGED interest. Justin took out a student loan for his first year of college. He borrowed $6,000 for 4 years. He was charged an interest rate of 4.5%. How much simple interest will Justin have to pay back on his loan?

Use the formula $d = rt$ to evaluate distances.

7. Find the distance traveled when r = 10 miles per hour and t = 4 hours.

8. Irina drives at a speed of 55 miles per hour. She drives for 3 hours before having to stop for gas. How far has she traveled so far?

9. Matthew ran 9 miles per hour for 1.5 hours. How far did he run?

10. Kirsten walked for 30 minutes at a speed of 5 miles per hour. How many miles did she walk?

11. The Gonzalez family lives in Milton-Freewater, Oregon. They drove to Denver, Colorado for summer vacation. On the first day they drove 9 hours at an average speed of 62 miles per hour. They stopped for the night in Salt Lake City. The next day they drove on to Denver at an average speed of 55 miles per hour for 9.75 hours.

    a. How far did the Gonzalez family travel on the first day?

    b. What is the total distance traveled by the Gonzalez family from Milton-Freewater to Denver AND BACK? Show all work necessary to justify your answer.

Use the formula $B = \frac{h}{a}$ to find batting averages. Round to the nearest thousandth.

12. Find the batting average when $h = 16$ and $a = 43$.

13. Willie has been up to bat 74 times this season. He has 21 hits. What is his batting average?

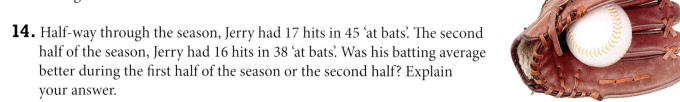

14. Half-way through the season, Jerry had 17 hits in 45 'at bats'. The second half of the season, Jerry had 16 hits in 38 'at bats'. Was his batting average better during the first half of the season or the second half? Explain your answer.

15. In 2012, Dustin Pedroia of the Boston Red Sox had 563 'at bats'. He had 163 hits during the season. What was his batting average?

16. Brad's dad drove 35 miles in 30 minutes. Brad determined his dad drove $1.\overline{16}$ miles per hour.

    a. Does Brad's calculation seem reasonable? Explain your reasoning.

    b. Describe the error in Brad's calculation and fix it.

17. Find another formula that is used in a work place that is not listed in this Block. What would the formula be used for?

REVIEW

Use the order of operations to evaluate.

18. $25 \div 5 + 7 \cdot 2$

19. $(27 - 3) \div 3 \cdot 2$

20. $(2 + 4)^2 - 10$

21. $3 \cdot 7 + 5^2 - 16$

22. $\frac{4 + 28}{6 - 2}$

23. $\frac{20(7 - 3)}{2 \cdot 5} - 7$

Evaluate each area, perimeter or circumference using the geometric formulas below.

Area = $\frac{1}{2}bh$

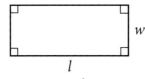
Area = lw
Perimeter = $2l + 2w$

Area = πr^2
Circumference = $2\pi r$

24. area of a circle when $r = 5$ m (Use 3.14 for π.)

25. area of a rectangle when $l = 6$ and $w = 9$

26. perimeter of a rectangle when $l = 10$ cm and $w = 4$ cm

27. area of a triangle when $b = 28$ ft and $h = 5$ ft

28. circumference of a circle when $r = 9$ (Use 3.14 for π.)

TIC-TAC-TOE ~ DREAM SCHOOL

Your town is considering building a new elementary school. The school board asked for possible floor plans to consider. Draw a "dream school" floor plan that meets the following criteria.

1. The school can only be one story.

2. Each room in the school should be triangular, rectangular or circular. You may also create rooms in your school that are any combination of those shapes.

Example:
Library with Bay Window

3. The school must contain one classroom for each grade, kindergarten through fifth grade. You should also include a gym, cafeteria and library. Include any additional rooms you would like.

4. Create a scale which you will use for your floor plan. Let 1 inch equal a set amount of feet. Check with an adult to determine if this is a reasonably sized school.

5. Draw your floor plan on a blank sheet of paper using a ruler. Write the real-life dimensions for each room on your floor plan. Record your scale on your building plan.

6. On a separate sheet of paper, determine the area (or square footage) of each room as they would be built in real-life.

7. Determine the overall square footage of your dream school.

TIC-TAC-TOE ~ CHILDREN'S BOOK

Math is used in many real-world experiences. Create a children's book that incorporates the concept of variables and at least one of the formulas listed below. Your book should have a cover, illustrations and a story line that is appropriate for children.

Use the simple interest formula $I = prt$ to evaluate the amount of interest earned.
Use the formula $d = rt$ to evaluate distances.
Use the formula $B = \frac{h}{a}$ to evaluate batting averages.

TIC-TAC-TOE ~ BANK INTEREST

Many financial calculations involve interest. Interest is the money you pay to the bank (in the case of a loan) or the bank pays you (in the case of a savings account). In **Lesson 2.4**, you learned to use the simple interest formula $I = prt$ where I is the interest, p is the principal, r is the interest rate per year and t is the time in years. Most banks do not use simple interest for savings accounts. They use a type of interest called compound interest.

Simple interest only earns interest based on the original deposit made. With compound interest, you earn interest on the original deposit as well as on the interest you have already received. To calculate *annual* (or yearly) compound interest, use the formula $A = p(1 + r)^t$ where A represents the ending balance in the account.

Example: Susan invested $2,000 in an account where interest is compounded yearly. She leaves the money in the account which earns 4% interest. How much will she have in five years?

$$p = 2,000 \qquad\qquad A = 2,000(1 + 0.04)^5$$
$$r = 0.04 \qquad\qquad A = 2,000(1.04)^5$$
$$t = 5 \qquad\qquad A \approx \$2,433.31$$

Find the total amount of money in an account that is compounded annually based on the information below.

1. $1,000 for three years in an account earning 3% interest
2. $8,000 for ten years in an account earning 4.5% interest
3. $500 in an account earning 7% interest for five years

Pretend you have been given $5,000 to place in a savings account for six years.

4. Research a minimum of four banks to determine the current savings account interest rate. Print a copy of the bank's web site or attach a brochure from the bank that shows the current interest rate.
5. Determine how much you would have in six years if you invested your money in each bank (assume the bank compounds annually).
6. At which bank would you open up a savings account?
7. What is the total difference, in dollars, of the best option compared to the worst option?

SIMPLIFYING ALGEBRAIC EXPRESSIONS

LESSON 2.5

 Recognize and combine like terms to simplify algebraic expressions.

Every algebraic expression has at least one **term**. A term is a number or is the product of a number and a variable. Terms are separated by addition and subtraction signs. A **constant** is a term that has no variable.

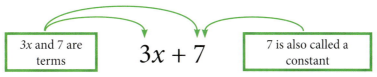

3x and 7 are terms
$$3x + 7$$
7 is also called a constant

Algebraic expressions often have **like terms** that can be combined. Like terms are terms that have the same variable. All constants are like terms. The **coefficient**, which is the number multiplied by a variable in a term, does not need to be the same in order for the terms to be like terms.

EXAMPLE 1

Match pairs of like terms.

6x	2y	3m	x
9	3m	5	14y

x can also be written $1x$

SOLUTION

There are four pairs of like terms. Each pair of like terms has the same variable or no variable.

$6x$ and x $2y$ and $14y$

9 and 5 $3m$ and $3m$

In order to simplify an algebraic expression you must combine all like terms. When combining like terms you must remember that the operation in front of the term (addition or subtraction) must stay attached to the term. Rewrite the expression by grouping like terms together before adding or subtracting the coefficients to simplify.

EXAMPLE 2

Simplify each algebraic expression by combining like terms.
a. $4x + 3x + 5x$
b. $8y + 6 + 3 + 4y + 1$
c. $9m + 10p - 3p - 2m + 4m$
d. $4d + 12 + d - 12$

SOLUTIONS

a. All terms are alike so add the coefficients together.

$$4x + 3x + 5x = 12x$$

b. There are two types of like terms in this algebraic expression. There are constants and terms that have the y variable. Rewrite the expression so the like terms are next to each other.

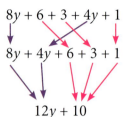

Add the coefficients of the like terms together.

$$12y + 10$$

c. Group the like terms together. The addition or subtraction sign must stay attached to the term.

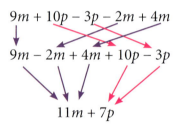

Add or subtract the coefficients of the like terms.

$$11m + 7p$$

d. Group like terms together. If a variable does not have a coefficient written in front of it, then the coefficient is 1.

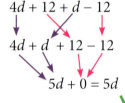

Add or subtract the coefficients of the like terms.

$$5d + 0 = 5d$$

Zero does not need to be written.

EXERCISES

Simplify each algebraic expression by combining like terms.

1. $4y + 2y + 3y$

2. $8x + x - 5x$

3. $10m - 4m + 2m - 3m$

4. $5x + 9y + 3x - 2y$

5. $14p + 8 + 1 - 14p$

6. $11 + 5d - 3d - 4$

7. $22u - 6u + 4t + 4t - 8u$

8. $x + y + x + x + 2y - x$

9. $15 + 8n + 3n - 2 - 13$

10. $23m + 4n - 3m$

11. $5 + 13 + 4f - f + 1$

12. $2x + 1 - x + 6x - 1$

13. $4g + 5k + 2g + 8h + k + 2h$

14. $3 + 10y + 14 - 10y - 17$

15. $a + 4b + c - a + 3c - 3b$

16. Let each apple $= x$ and each orange $= y$.

    a. Write an algebraic expression that represents the following diagram.

    b. Write a four-term algebraic expression for the following diagram:

    c. Simplify the algebraic expression from **part b**. How many terms does this expression have?

17. In your own words explain how you can tell when terms are like terms.

18. Write an algebraic expression that has terms that can be combined. Simplify your algebraic expression.

19. Uma thinks that $4x + 3 + x + 5$ is equal to $4x + 8$. Is she correct? Explain your reasoning.

20. Xavier said no terms could be combined in the expression shown below. Do you agree or disagree? Explain your reasoning.

$$5y + 6b + 5 + 6p$$

REVIEW

Evaluate each algebraic expression when $x = 3$.

21. $2x - 1$

22. $\dfrac{2x + 2}{9}$

23. $\frac{1}{3}x + 5$

24. $(x + 3)^2$

25. $(x + 2)^2$

26. $20 - 5x$

27. $\dfrac{x}{3} + 2$

28. $4(x + 7)$

29. $\dfrac{x}{6} + \dfrac{1}{3}$

TIC-TAC-TOE ~ POPULATION DENSITY

The human population of different areas of the world can be studied by examining population density. Population density most often describes the number of people per square mile. You can calculate population density by using a formula.

$$\text{Population Density} = \frac{\text{population}}{\text{area in square miles}}$$

Example: Iowa's population in 2011 was about 3,000,000. The state of Iowa is approximately 56,000 square miles. The population density of Iowa in 2011 was:

$$\text{Population Density} = \frac{3000000}{56000} \approx 54 \text{ people per square mile.}$$

1. Approximately 38 million people lived in the State of California in 2011. California is approximately 156,000 square miles. Find the population density.

2. About 128 million people live in Japan. The entire country of Japan is only 146,000 square miles. Find the population density of Japan.

3. One of the most populated areas in the United States is Washington DC. The size of Washington DC is approximately 61 square miles. The population of Washington DC was 618,000 in the year 2011. What was the population density of Washington DC in 2011? Would you like to live in a place this densely populated?

4. Which state do you think has the lowest population density? Research that state to find the most recent population count and the size of the state in square miles. Calculate the population density.

5. There are approximately 7 billion humans on earth. There are 58 million square miles on Earth where humans can live. What is the population density of the Earth?

6. Find the population and size (in square miles) of your county on the internet. What is the population density of your county?

7. Do you think the population density of your county will increase or decrease in the next ten years? Explain your answer.

THE DISTRIBUTIVE PROPERTY

 Use the Distributive Property to perform calculations.

Davis High School sold coupon books as a fundraiser. Each coupon book cost $12. One student sold 8 coupon books. Another student sold 14 coupon books.

To figure out how much money these two students brought in for the fundraiser, multiply the cost per coupon book times the total number of books sold.

METHOD 1	METHOD 2
Cost per book × total number of books sold = $12 (8 + 14) = $12(22) = $264	Cost per book × books sold by Student #1 **+** Cost per book × books sold by Student #2 $12 · 8 = $96 + $12 · 14 = $168 TOTAL = $264

The two methods for solving the coupon book example illustrate different ways to solve a problem. Another way to organize and solve this example is to use the **Distributive Property**. The Distributive Property is an important property that allows you to simplify computations or algebraic expressions that include parentheses.

> **THE DISTRIBUTIVE PROPERTY**
> For any numbers a, b and c:
> $a(b + c) = a(b) + a(c)$ or $ab + ac$
> $a(b − c) = a(b) − a(c)$ or $ab − ac$

EXAMPLE 1

Rewrite each expression using the Distributive Property. Evaluate.
a. 12(8 + 14)   b. 6(8 − 2)   c. 3(2.2 + 4)

SOLUTIONS

a. $12(8 + 14) = 12(8) + 12(14) = 96 + 168 = 264$

b. $6(8 − 2) = 6(8) − 6(2) = 48 − 12 = 36$

c. $3(2.2 + 4) = 3(2.2) + 3(4) = 6.6 + 12 = 18.6$

> This solution shows how to solve the coupon book problem above using the Distributive Property.

The Distributive Property is very useful when doing mental math calculations. Certain numbers are easier to multiply together than others. In the next example, notice how you can rewrite a number as a sum or difference of two other numbers. If you choose numbers that are easier to work with, you will be able to do the math mentally.

EXAMPLE 2

Find the product by using the Distributive Property and mental math.
a. 3(102)   b. 5(197)   c. 6(4.1)

SOLUTIONS

a. Rewrite 102 as 100 + 2.
   Distribute.
   Add.

$$3(100 + 2)$$
$$3(100 + 2) = 3(100) + 3(2)$$
$$= 300 + 6$$
$$= 306$$

b. Rewrite 197 as 200 − 3.
   Distribute.
   Subtract.

$$5(200 - 3)$$
$$5(200 - 3) = 5(200) - 5(3)$$
$$= 1000 - 15$$
$$= 985$$

c. Rewrite 4.1 as 4 + 0.1.
   Distribute.
   Add.

$$6(4 + 0.1)$$
$$6(4 + 0.1) = 6(4) + 6(0.1)$$
$$= 24 + 0.6$$
$$= 24.6$$

EXPLORE!   **SHOPPING SPREE**

You just won a $5,000 shopping spree at the mall. Determine if you have any money left after purchasing items for you and your friends during the first day of your shopping spree.

Step 1: You purchased two tablets for $396 each. Use the Distributive Property to rewrite the expression. Choose whether to use + or − between the numbers in the parentheses. Evaluate your expression.

$$2(\$396) = 2(\underline{\quad} \pm \underline{\quad})$$

Step 2: You also purchased eight shirts for $9.25 each. Use the Distributive Property to rewrite the expression. Evaluate the expression.

$$8(\$9.25) = 8(\underline{\quad} \pm \underline{\quad})$$

Step 3: Next, you purchased seven bottles of perfume for $104 each. Use the Distributive Property to rewrite the expression. Evaluate the expression.

$$7(\$104) = 7(\underline{\quad} \pm \underline{\quad})$$

Step 4: Finally, you splurged and bought three flat-screen televisions at a cost of $997 each. Use the Distributive Property to rewrite the expression. Evaluate the expression.

$$3(\$997) = 3(\underline{\quad} \pm \underline{\quad})$$

Step 5: Find the amount of money you have spent. How much money is left of your $5,000 shopping spree?

EXERCISES

Copy each statement. Fill in the blanks using the Distributive Property.

1. $4(5 + 7) = 4(\_\_\_\_) + \_\_\_\_(7)$

2. $8(12 - 9) = \_\_\_\_(12) - \_\_\_\_(9)$

3. $5(6 + 2) = 5(\_\_\_\_) + \_\_\_\_(\_\_\_\_)$

4. $\frac{1}{2}(11 + 4) = \frac{1}{2}(\_\_\_\_) + \_\_\_\_(4)$

5. Jackson works during the summer as a babysitter for the neighbor kids. He earns $6 for each hour he babysits. On the first day of summer, he works for 4 hours. He babysits for 7 hours on the second day.

a. Explain in words how to determine the total amount of money Jackson made during his first two days of babysitting.

b. Write an expression and evaluate the total amount of money Jackson earned in two days.

Rewrite each expression using the Distributive Property. Evaluate each expression.

6. $5(6 + 8)$

7. $3(20 + 4)$

8. $6(30 - 1)$

9. $7(6 - 0.2)$

10. $9(3 + 1.2)$

11. $2(60 + 8)$

12. $12(3 + 0.1)$

13. $5(6 + 0.3)$

14. $11(9 - 4)$

15. Explain how the Distributive Property helps in using mental math to find a product.

Find the product by using the Distributive Property.

16. $4(201)$

17. $7(99)$

18. $3(9.2)$

19. $5(1003)$

20. $2(12.98)$

21. $8(305)$

22. Shasta finds six DVDs she wants to purchase at the store. Each DVD costs $14.95.

a. Show how Shasta could use the Distributive Property to help mentally calculate the total cost of the DVDs.

b. How much will she pay for the six DVDs?

23. The formula $P = 2(l + w)$ can be used to find the perimeter of a rectangle. Find the perimeter of a rectangular garden with a length of 20 feet and a width of 16 feet.

Use the Distributive Property to evaluate.

24. $4(5 + 8) + 3(10 + 2)$

25. $3(10 - 3) + 8(20 + 1)$

26. Drew went to the grocery store with $20. He mentally calculated his total as he put items in his cart so that he would not overspend. Compute the value of the items he put in the cart using the Distributive Property.

    a. first item: 5 cans of soup for $0.95 each

    b. second item: 3 frozen pizzas for $3.09 each

    c. third item: 4 king-sized candy bars for $1.05 each

    d. Did Drew have enough money to pay for all the items listed above? Use mathematics to justify your answer.

27. Susan's entire family took the train to watch a baseball game. The admission to the game was $8.00 per person. The round-trip pass for the train cost $4.25 per person. There are 6 people in Susan's family. Show two ways to find the total cost of the outing.

28. Jamal decided to break a number into three parts before doing the Distributive Property. Does his method work? Support your answer with words and/or symbols.

$$4(75) = 4(25 + 25 + 25)$$
$$= 100 + 100 + 100$$
$$= 300$$

29. Find the value of the following expression: $4(8 \times 10) + 5(7 - 1) + 3(11 + 2) - 6(8 + 5)$.

REVIEW

Evaluate each expression when $x = 2$.

30. $5(x + 1.4)$

31. $(x + 5)^2 + 2x$

32. $\frac{3x - 2}{6}$

33. $16\left(\frac{1}{2}x + 1\right)$

Use the simple interest formula, $I = prt$, to evaluate the amount of interest earned.

34. Find the amount of interest earned when $p = \$200$, $r = 2\%$ and $t = 12$ years.

35. Find the amount of interest earned when $p = \$10,000$, $r = 6\%$ and $t = 2$ years.

36. Peter deposited $400 in an account for 1 year at 7% interest. How much money did he earn?

37. Lindsey deposited $150 in an account for 3 years with an interest rate of 10%. J.R. deposited $400 in an account for 3 years with an interest rate of 4%.

    a. Predict who will have made more in interest at the end of three years. Explain your reasoning.

    b. Was your prediction correct? Use words and/or numbers to show how you determined your answer.

TIC-TAC-TOE ~ GROCERY SHOPPING

The Distributive Property can help you calculate the cost of items mentally. Prices at grocery stores are rarely in whole dollar amounts, except for sale items. To do this activity, take a trip to a grocery store. Remember that you DO NOT need to buy the items on the list, only find the prices.

1. Copy the following table and take it to a local grocery store.

2. Record the normal price per item. DO NOT USE SALE PRICES.

3. Determine whether you will use the Distributive Property to calculate the expenses for each item. Write *yes or no* in the "Use the Distributive Property" column. Use the Distributive Property for at least 3 items.

4. If you use the Distributive Property, record your expression in the "calculations" column. If you do not use the Distributive Property, record the operations used.

5. Determine the total cost for each item on the list.

6. Determine the total cost of the shopping trip.

7. Explain how you chose when to use the Distributive Property and when not to.

GROCERY STORE:				
Items	Cost Per Item	Use the Distributive Property?	Calculations	Total Cost
3 Boxes of Macaroni & Cheese				
7 Cans of Chili				
6 Containers of Ice Cream				
5 Gallons of Milk				
9 Bags of Pretzels				

USING THE DISTRIBUTIVE PROPERTY WITH VARIABLES

LESSON 2.7

Simplify expressions with variables.

Many algebraic expressions have parentheses. In order to simplify algebraic expressions with parentheses, the Distributive Property must be used. Multiply the front factor by each term in the parentheses.

EXAMPLE 1

Use the Distributive Property to simplify each expression.
a. $2(x + 6)$ **b.** $5(y - 8)$ **c.** $4(3x - 2)$

SOLUTIONS

a. $2(x + 6) = 2(x) + 2(6) = 2x + 12$

b. $5(y - 8) = 5(y) - 5(8) = 5y - 40$

> Only multiply the two coefficients together.
> Think of it as 4 groups of $3x$.

c. $4(3x - 2) = 4(3x) - 4(2) = 12x - 8$

Algebraic expressions or equations often have like terms that can be combined. If there are parentheses involved in the expression, the Distributive Property must be used FIRST before combining like terms.

EXAMPLE 2

Simplify by distributing and combining like terms.
$5(x + 4) + 3x + 3$

SOLUTIONS

Distribute first. $5(x + 4) + 3x + 3 = 5x + 20 + 3x + 3$

Use the Commutative Property $5x + 3x + 20 + 3$
to group like terms.

Combine like terms. $8x + 23$

$5(x + 4) + 3x + 3 = 8x + 23$

Two algebraic expressions are equivalent expressions if they represent the same simplified algebraic expression.

In this matching game, each card in the deck has an algebraic expression on it. The goal of the game is to match the six LETTER Expression Cards (A, B, C...) to their equivalent NUMBER Expression Cards (1, 2, 3...).

Step 1: Simplify each LETTER Expression Card.

Step 2: Match each LETTER Expression Card to its equivalent NUMBER Expression Card.

Step 3: Create two more LETTER Expression Cards and corresponding NUMBER Expression Cards to be used in a future matching game.

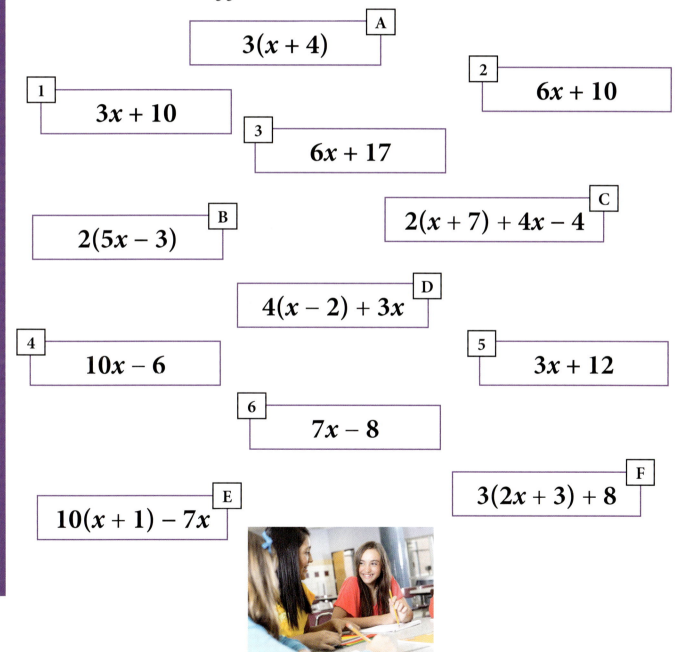

A

$$3(x + 4)$$

2

$$6x + 10$$

1

$$3x + 10$$

3

$$6x + 17$$

B

$$2(5x - 3)$$

C

$$2(x + 7) + 4x - 4$$

D

$$4(x - 2) + 3x$$

4

$$10x - 6$$

5

$$3x + 12$$

6

$$7x - 8$$

F

$$3(2x + 3) + 8$$

E

$$10(x + 1) - 7x$$

EXERCISES

Use the Distributive Property to simplify.

1. $5(x + 7)$

2. $3(y - 4)$

3. $6(m + 1)$

4. $7(x - 0.1)$

5. $4(3p + 2)$

6. $2(5x - 6)$

7. $3(20 - x)$

8. $5(2y + 10)$

9. $11(2m - 4)$

10. Use words to explain the steps needed to simplify the following expression.
$$3(x + 6) + 2x + 1$$

11. Write two equivalent expressions. Explain or show how you know the two expressions are equivalent.

Simplify each expression.

12. $6(x + 2) + 4x$

13. $2(3x + 5) - 7$

14. $3(x - 10) - x$

15. $10(x - 2) - 5x$

16. $2(5x + 2) + 3x + 5$

17. $5(x + 4) + x - 8$

18. $3 + 2x + 4(x + 1)$

19. $7(2x + 3) - 10$

Write and simplify an expression for the perimeter of each figure.

20.

4x   5x

3x

21.

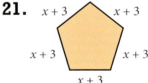

x + 3   x + 3

x + 3   x + 3

x + 3

22.

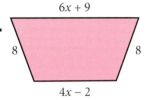

6x + 9

8   8

4x − 2

23.

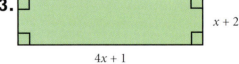

x + 2

4x + 1

Write and simplify an expression for the area of each figure.

24.

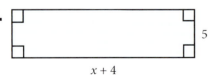

5

x + 4

25.

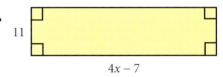

11

4x − 7

In each set of three expressions, two are equivalent. Simplify each expression to find the two equivalent expressions.

26. i. $2(x + 7)$
ii. $2(x + 4)$
iii. $2(x + 1) + 6$

27. i. $3(2x + 1)$
ii. $6(x + 1)$
iii. $6(x + 2) - 9$

Use the Distributive Property to simplify.

28. $4(x + 5) + 3(x + 2)$

29. $3(x + 8) + 4(x - 1)$

30. Jeremy states that $4(x + 2)$ is equivalent to $4(2 + x)$. Do you agree with him? Why or why not?

REVIEW

Write an algebraic expression for each phrase.

31. the sum of x and twelve

32. six times x

33. the product of x and seven

34. the quotient of x divided by five

35. Evaluate **Exercises 31–34** when $x = 25$.

36. A cheetah traveling at full speed covers 1.2 miles per minute.
    a. How far will the cheetah travel in 3 minutes?
    b. How far will the cheetah travel in 8 minutes?
    c. How far will the cheetah travel in m minutes?

TIC-TAC-TOE ~ EXPRESSIONS GAME

Create a memory card game using equivalent expressions. Create pairs of cards that have expressions that, when simplified, are equivalent. The cards should be made on thicker paper (such as card stock, construction paper, index cards or poster board). The game must have a minimum of 24 cards.

Below is an example of two cards with equivalent expressions.

$$3x + 7 + 5x + 2$$ $$8x + 9$$

Play the game with a friend, parent or classmate. List each player's name on a sheet of paper. Record each pair of equivalent expressions under the name of the player who won each set.

Vocabulary

algebraic expression	Distributive Property	like terms
coefficient	equivalent expressions	term
constant	evaluate	variable
	formula	

Write expressions involving variables.
Evaluate algebraic expressions.
Use geometric formulas to find area, perimeter and volume.
Use formulas in a variety of situations.
Recognize and combine like terms to simplify algebraic expressions.
Use the Distributive Property to perform calculations.
Simplify expressions with variables.

Lesson 2.1 ~ Variables and Expressions

Write an algebraic expression for each phrase.

1. nine more than d

2. one subtracted from p

3. forty less than w

4. twenty-two divided by m

5. a number y decreased by ten

6. three times a number x

Write a phrase for each algebraic expression.

7. $6y$

8. $d - 9$

9. $12 \div x$

10. $m + 3$

11. Gina's Candy Shack sells fudge and caramel apples. A piece of fudge costs $0.75 less than a caramel apple.

    a. A caramel apple costs $2.75. How much does a piece of fudge cost?

    b. A caramel apple costs y dollars. What algebraic expression represents the cost of a piece of fudge? Explain how you know your expression is correct.

Evaluate each expression.

12. $x - 6$ when $x = 29$

13. $5b + c$ when $b = 4$ and $c = 2$

14. $3y + 2$ when $y = 5$

15. $42 - 2x$ when $x = 20$

16. $\frac{1}{2}m + 6$ when $m = 10$

17. $\frac{w}{7}$ when $w = 7$

Copy each table. Complete each table by evaluating the given expression for the values listed.

18.

x	$5x + 2$
0	
1	
4	
7	

19.

x	$\frac{1}{2}x - 1$
2	
6	
9	
14	

20. Elderberry Elementary School is having a hot dog feed. The staff needs to purchase packages of hot dogs and buns. Each package of hot dogs costs $2.00 and the packages of buns cost $1.50 each. The staff uses the algebraic expression $2.00x + 1.50y$ to calculate their total expenses.

    a. What does the x variable stand for? y variable?

    b. They end up buying 15 packages of hot dogs and 20 packages of buns. How much will they spend altogether?

Lesson 2.3 ~ Evaluating Geometric Formulas

Evaluate each area, perimeter or circumference using the geometric formulas below.

Area $= \frac{1}{2}bh$

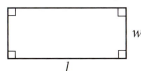

Area $= lw$
Perimeter $= 2l + 2w$

Area $= \pi r^2$
Circumference $= 2\pi r$

21. area of a triangle when $b = 10$ *in* and $h = 4$ *in*

22. area of a circle when $r = 2$ *m* (Use 3.14 for π.)

23. perimeter of a rectangle when $l = 14$ and $w = 5$

24. area of a rectangle when $l = 7$ *ft* and $w = 6$ *ft*

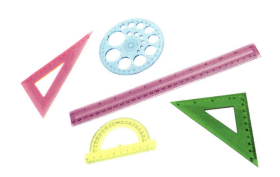

25. circumference of a circle when $r = 4$ (Use 3.14 for π.)

Evaluate each surface area or volume of a rectangular prism using the geometric formulas below.

Volume = lwh
Surface Area = $2(lw + wh + hl)$

26. volume of a rectangular prism when $l = 6$, $w = 2$ and $h = 5$

27. surface area of a rectangular prism when $l = 2$ cm, $w = 1$ cm and $h = 7$ cm

Lesson 2.4 ~ Evaluating More Formulas

Use the simple interest formula, $I = prt$, to evaluate the amount of interest earned.

28. Find the amount of interest earned when $p = \$1,000$, $r = 5\%$ and $t = 4$ years.

29. Abe put $100 in an account for one year at 6% interest. How much interest did he earn?

30. When Kisha was born, her parents deposited $2,000 in an account for college. The money was there for 18 years at 8% interest. Use the simple interest formula. How much interest did Kisha's college account earn in 18 years? Show all work necessary to justify your answer.

Use the formula $d = rt$ to evaluate distances.

31. Find the distance traveled when $r = 8$ miles per hour and $t = 5$ hours.

32. Clint drives at a speed of 62 miles per hour. He drives 4 hours before stopping at a rest stop. How far has he traveled?

Use the formula $B = \frac{h}{a}$ to evaluate batting averages. Round to the nearest thousandth.

33. Find the batting average when $h = 28$ and $a = 88$.

34. Zurina has been up to bat 25 times this season. She has 12 hits. What is her batting average?

Lesson 2.5 ~ Simplifying Algebraic Expressions

Simplify each algebraic expression by combining like terms.

35. $8x + 2x + 5x$

36. $3y + 2 + 6y + 1$

37. $11m - 2m + m$

38. $4x + 3y + x - 2y$

39. $5y + 10 + 4y - 10$

40. $x + y + x + x - x - y$

Rewrite each expression using the Distributive Property and evaluate.

41. $3(20 + 2)$

42. $7(8 + 3)$

43. $4(5 - 0.1)$

44. $10(9 + 0.3)$

Find the product by using the Distributive Property.

45. $5(201)$

46. $8(98)$

47. $4(2005)$

48. $7(10.9)$

49. Melanie buys 8 bags of trail mix. Each bag costs $1.97.
    a. Show how Melanie could use the Distributive Property to
       mentally calculate the total cost of the eight bags of trail mix.
    b. How much will she pay for all 8 bags of trail mix?

Lesson 2.7 ~ Using the Distributive Property with Variables

Use the Distributive Property to simplify.

50. $7(x - 3)$

51. $4(x + 10)$

52. $6(2x + 5)$

Simplify each expression.

53. $6(x + 1) + 4x$

54. $2(9x + 10) + 4x - 7$

55. $4(x + 3) + 2x + 5$

56. $10(x - 2) - x$

In each set of three expressions, two are equivalent. Simplify each expression to find the two equivalent expressions.

57. **i.** $5(x + 2) + 4$
    ii. $5(x + 3)$
    iii. $5(x + 4) - 5$

58. **i.** $4(x + 6) - 4$
    ii. $2(2x + 10) + 10$
    iii. $6(x + 2) - 2x + 8$

CAREER FOCUS

CHRIS
DIGITAL ARCHIVIST

I am a digital archivist in the library of a University. My job is to create digital objects of historical documents to place on the web. I help oversee all of the scanning, formatting and description writing of the items that we put on the internet for people to see and use. I work every day with important documents and artifacts that are part of our school library. My job is to make sure that these items are accessible to as many people as possible.

I use math in the formatting and scanning of materials that go onto websites. Geometry helps me determine the sizes and shapes of the items that will go on a page. I also use percentages and other mathematical calculations to determine how to get the best quality of photo into a format that people will be able to see. There are many different types of computers and monitors that people use. Math helps me make sure that my websites will look good on any type of equipment.

Most people who want to be digital archivists need to go to college to get an undergraduate degree. After that, they usually need to get a Master's degree in library science. To get both degrees takes about six years. There are now many schools in the United States that offer programs for digital archiving.

A good starting salary in this career would be around $40,000 per year. Most archivists start at a salary around $30,000 per year. Digital archivists can eventually earn close to $80,000 per year. There are extra skills you need to have to become a digital archivist. Because of the extra skills, digital archivists tend to make a bit more than other archivists.

There are two things about my job I find especially rewarding. One is the opportunity to work with the writings of historical figures on a daily basis. The second is making these priceless materials available to a wide audience through the digital products I help create.

CORE FOCUS ON INTRODUCTORY ALGEBRA
BLOCK 3 ~ SOLVING EQUATIONS

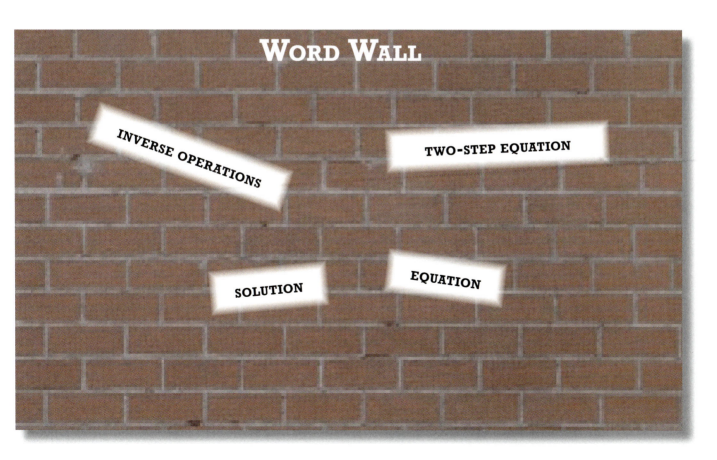

WORD WALL

INVERSE OPERATIONS

TWO-STEP EQUATION

SOLUTION

EQUATION

BLOCK 3 ~ SOLVING EQUATIONS
TIC-TAC-TOE

LIKE TERMS

Learn how to solve two types of equations with like terms.

See page 88 for details.

EQUATION BOARD GAME

Create a board game where players move toward the goal by solving equations.

See page 106 for details.

FLIP BOOK

Make a flip book to help tutor students who are trying to learn how to solve one-step equations.

See page 83 for details.

GOING TO THE COUNTY FAIR

Find the number of children and adults who went to a County Fair.

See page 76 for details.

MISSING DIMENSIONS

Find the values of missing dimensions in complex geometry figures.

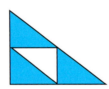

See page 100 for details.

EQUATION BINGO

Design an equation BINGO game. Players solve equations correctly in order to win.

See page 92 for details.

EARNING AN 'A' IN MATH

Look at four students' test scores to determine how each student must do on the final test to earn an 'A'.

See page 101 for details.

MISSING PRICE TAGS

Use the percent equation to determine original prices.

See page 87 for details.

DESIGN A WORKOUT

Use a calorie-burning chart to design different workouts.

See page 80 for details.

EQUATIONS AND SOLUTIONS

LESSON 3.1

> Determine if a number is a solution of an equation.

An **equation** is a mathematical sentence that contains an equals sign (=) between two expressions.

Equations involving a variable are neither true nor false until the equation is evaluated with a given value for the variable. A value is considered the **solution** of an equation if it makes the equation true.

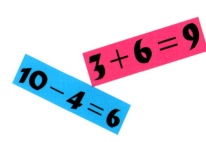

$$x + 7 = 20$$
$$13 + 7 = 20$$
$$20 = 20$$

The value of x that makes this equation true is 13. Thirteen is the solution of the equation.

The equation.

The equation is true.

EXAMPLE 1

Determine if the number given is the solution of the equation.

a. $x - 5 = 12$ *Is 7 the solution?*

b. $8y = 32$ *Is 4 the solution?*

c. $\frac{m}{3} = 9$ *Is 27 the solution?*

SOLUTIONS

a. Substitute the given value for the variable into the equation to see if it makes the equation true.

$$x - 5 = 12$$
$$7 - 5 \overset{?}{=} 12$$
$$2 \neq 12 \quad \textbf{NOT A SOLUTION.}$$

The $\neq$ symbol shows that one side of the expression is "not equal to" the other side of the expression.

b. Substitute 4 for y to see if it makes the equation true.

$$8y = 32$$
$$8(4) \overset{?}{=} 32$$
$$32 = 32 \quad \textbf{A SOLUTION.}$$

c. Substitute 27 for m to see if it makes the equation true.

$$\frac{m}{3} = 9$$
$$\frac{27}{3} \overset{?}{=} 9$$
$$9 = 9 \quad \textbf{A SOLUTION.}$$

EXAMPLE 2

Identify whether each number in the table is a solution of the equation.

x	$14 - x = 8$	Solution?
4		
6		
9		

SOLUTION

Each value must be substituted for the variable to determine if it is a solution. Remember that the number is only a solution if it makes the equation true.

x	$14 - x = 8$	Solution?
4	$14 - 4 \overset{?}{=} 8$ $10 \neq 8$	NO
6	$14 - 6 \overset{?}{=} 8$ $8 = 8$	YES
9	$14 - 9 \overset{?}{=} 8$ $5 \neq 8$	NO

Many situations in life require you to check if a solution actually makes a statement true.

A female polar bear at the zoo weighs 120 pounds less than the zebra in a nearby exhibit. The polar bear weighs 570 pounds. Elijah says the weight of the zebra is 450 pounds. Katrina says the zebra weighs 690 pounds. Who is correct?

Write the equation that represents the original statement:

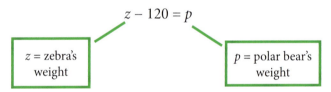

$$z - 120 = p$$

z = zebra's weight

p = polar bear's weight

The polar bear weighs 570 pounds. Substitute this value for p.

$$z - 120 = 570$$

Determine the weight of the zebra by checking the two possible solutions.

$$450 - 120 \overset{?}{=} 570$$
$$330 \neq 570 \qquad \textbf{NO}$$

$$690 - 120 \overset{?}{=} 570$$
$$570 = 570 \qquad \textbf{YES}$$

The zebra weighs 690 pounds. Katrina was correct.

EXERCISES

Determine if the number given is the solution of the equation.

1. $x - 6 = 3$ *Is 7 the solution?*

2. $6y = 54$ *Is 9 the solution?*

3. $w + 8 = 22$ *Is 14 the solution?*

4. $h - 14 = 4$ *Is 26 the solution?*

5. $\frac{p}{11} = 6$ *Is 66 the solution?*

6. $7x = 21$ *Is 4 the solution?*

7. $6 + k = 7.2$ *Is 2.2 the solution?*

8. $y - \frac{1}{2} = 4\frac{1}{2}$ *Is 5 the solution?*

Copy and complete each table. Determine if each number in the table is a solution of the equation.

9.

x	$9 - x = 3$	Solution?
3		
6		
9		

10.

x	$x + 11 = 24$	Solution?
11		
12		
13		

11.

x	$4x = 4.8$	Solution?
1		
1.2		
1.8		

12.

x	$\frac{x}{6} = 7$	Solution?
42		
48		
60		

13. Patrick did not put any effort into the "Equations Quiz" his teacher gave on Friday. He used the answer of 6 as the solution to all five problems on the quiz.

    a. Which problem(s) did he get correct with this guess?
    b. What percent of the questions did he get correct?
    c. Was Patrick's strategy a good method to use? Why or why not?

> **Equations Quiz**
>
> 1. $x + 4 = 9$
> 2. $20 - y = 14$
> 3. $5m = 45$
> 4. $t - 4 = 7$
> 5. $\frac{b}{2} = 12$

14. The distance around the bases of an official baseball diamond is 360 feet. Sherilynn decided to let the variable, b, represent the distance from one base to the next. She wrote the equation $4b = 360$ to represent the situation.

    a. How did Sherilynn come up with this equation?
    b. Sherilynn also stated that the solution to the equation is 90 feet. Is she right? Use words and/or numbers to show how you determined if she was correct.

Write an algebraic equation for each sentence. Use *x* as the variable. Match each equation to its solution.

15. A number plus six equals ten.

A. $x = 20$

16. A number minus fourteen equals three.

B. $x = 17$

17. Three times a number is twenty-seven.

C. $x = 9$

18. The quotient of a number divided by five is four.

D. $x = 3$

19. Eleven more than a number is fourteen.

E. $x = 4$

20. Two subtracted from a number equals eight.

F. $x = 10$

21. Zach uses a stair-stepping machine at the gym every day. The machine uses the formula $c = 7m$ to calculate the number of calories burned (*c*) when the stair-stepper is used for *m* minutes. At the end of the work-out, the machine indicates that Zach has burned 175 calories.

    a. Substitute 175 for *c* in the equation.
    b. Zach thought he was on the stair-stepper for 20 minutes. Is this the solution to the equation from **part a**?
    c. Was Zach on the stair-stepper for more or less than 20 minutes? Explain how you know your answer is correct.

22. Explain in words how you can determine if a given value is a solution to an equation.

23. Write four equations involving a variable, each using a different operation (add, subtract, multiply and divide), that are true when 5 is substituted for the variable. Use mathematics to justify your answer.

REVIEW

Simplify each expression.

24. $3(x + 5)$

25. $2(3x - 8)$

26. $4(x + 2) + 3x$

27. $20(x + 2) - 5$

28. $5(x + 1) + 2(x + 3)$

29. $4(2x + 3) + 3x - 10$

Evaluate each expression when *x* = 4, *y* = 20 and *z* = 0.

30. $x + y + z$

31. $2(y + x)$

32. $\frac{z}{x}$

33. $5 + y \div x$

34. $6(x + z) + y$

35. $6z + 3x$

SOLVING EQUATIONS USING MENTAL MATH

LESSON 3.2

 Solve one-step equations using mental math.

On "Lucky Number Thursday" Susan won 10 pounds of free produce at the local grocery store by having the lucky number on her store savings card. Susan filled a basket with fruits and vegetables. It weighed $8\frac{1}{2}$ pounds. Which additional item could Susan put in the basket to get exactly to the 10 pound limit?

Cantaloupe	$3\frac{1}{2}$ pounds
Watermelon	5 pounds
Pear	$\frac{1}{2}$ pound
Mango	$1\frac{1}{2}$ pounds
Squash	$2\frac{1}{2}$ pounds

In this lesson, you will learn a problem-solving strategy to solve equations using mental math. You will be given equations and asked to find the solution to each equation. The steps to mentally solve an equation are: Guess, Check and Revise.

Try this method on the grocery store example.

GUESS: Squash

CHECK: $8\frac{1}{2} + 2\frac{1}{2} = 11$ It is over the 10 pound limit.

REVISE: Mango

> After each revision you must check your new guess.

CHECK: $8\frac{1}{2} + 1\frac{1}{2} = 10$ Exactly 10 pounds → Mango is the correct item.

EXAMPLE 1

Use the Guess, Check and Revise method to solve each equation.
a. $y + 7 = 15$ b. $6x = 30$

SOLUTIONS

a. $y + 7 = 15$

Guess	→	Check	→	Revise	→	Check
$y = 9$		$9 + 7 \overset{?}{=} 15$		$y = 8$		$8 + 7 \overset{?}{=} 15$
		$16 \neq 15$				$15 = 15$

> $y = 8$ is the solution.

> Too high. Pick a lower number.

b. $6x = 30$

Guess	→	Check
$x = 5$		$6(5) \overset{?}{=} 30$
		$30 = 30$

> The first guess worked. $x = 5$

Some equations contain numbers that make it difficult to solve the equation using mental math. Estimation is a great strategy to use in these situations.

EXAMPLE 2

Two sisters, Kylie and Sara, went shopping for school clothes. Sara spent $105.11 and Kylie wasn't sure how much she spent. The sisters spent a total of $187.68. Write an equation and use it to estimate how much Kylie spent on clothes.

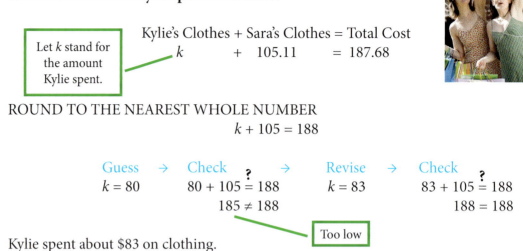

SOLUTION

Let k stand for the amount Kylie spent.

Kylie's Clothes + Sara's Clothes = Total Cost
$$k \quad + \quad 105.11 \quad = \quad 187.68$$

ROUND TO THE NEAREST WHOLE NUMBER
$$k + 105 = 188$$

Guess → Check _? → Revise → Check

Guess	Check	Revise	Check
$k = 80$	$80 + 105 \overset{?}{=} 188$	$k = 83$	$83 + 105 \overset{?}{=} 188$
	$185 \neq 188$		$188 = 188$

Too low

Kylie spent about $83 on clothing.

EXPLORE!

RECORD-SETTING
Source: *Guiness Book of World Records*

Step 1: Helen grew her fingernails out so each nail is 24.2 *cm* long. The longest fingernail on record is 76.7 *cm*. Write an addition equation and use it to estimate how much longer Helen's nails need to be to tie the record.

Step 2: John Evans balanced a 352 pound car on his head. Ryan worked his way up to balancing a 151 pound car on his head. Write an addition equation and use it to determine how many more pounds Ryan needs to balance on his head to tie the record.

Step 3: A Japanese pig named Kotetsu has the record for the highest jump by a pig. His jump measured 27.5 inches. An American pig jumped 12.2 inches. Write an addition equation and use it to estimate how many more inches the American pig needs to tie the record.

Step 4: Mac gathered 1,498 people for an Irish Dance. This is about 6,100 people fewer than the record set in Cork, Ireland on September 10, 2005. Write a subtraction equation and use it to estimate how many people attended the Irish Dance in Ireland in 2005.

EXERCISES

Use the Guess, Check and Revise method to solve each equation.

1. $3x = 18$

2. $y + 9 = 20$

3. $44 - h = 31$

4. $x - 6 = 13$

5. $20 \div m = 10$

6. $1.5p = 3$

7. $6 + y = 29$

8. $\frac{w}{7} = 3$

9. $4 + b = 4$

10. $15 - z = 1$

11. $c - 7 = 2$

12. $\frac{x}{8} = 6$

13. Shannon weighs 98 pounds. When she steps on the scale holding her cat, the scale reads 112 pounds. Let c represent the weight of the cat.
    a. Write an equation using the information given.
    b. Solve the equation using mental math. How much does the cat weigh?

14. Otis bought 12 bottles of soda pop. He spent a total of $18. Let p represent the price of one bottle of soda pop.
    a. Write a multiplication equation using the information given.
    b. How much does each bottle of soda pop cost? Use words and/or numbers to show how you determined your answer.

Write an algebraic equation for each sentence. Use x as the variable. Find the solution using mental math.

15. The sum of a number and six equals fifteen.

16. Four times a number is sixteen.

17. The quotient of a number divided by two is seven.

18. Ten less than a number is two.

19. The Coast Douglas Fir was considered the tallest tree in Oregon at 329 feet. The tallest tree was 94 feet taller than Haystack Rock off Cannon Beach. What is the height of Haystack Rock? Explain how you know your answer is correct.

Estimate the solution of each equation to the nearest whole number.

20. $4.1x = 28.2$

21. $y + 60.9 = 69.8$

22. $3\frac{1}{8} - m = 1\frac{1}{10}$

23. $x - 12.9 = 0$

24. $7.9 \div m = 4.05$

25. $p + 5.1 = 11.9$

26. $\frac{y}{5.1} = 2.98$

27. $12.4k = 0$

28. $18.1 - d = 12$

29. Tim and Tom went to a soccer game. It cost $23.80 for two tickets to the game. They also spent money on souvenirs. The equation $s + 23.80 = 78.12$ describes Tim and Tom's spending where s stands for the total amount of money spent on souvenirs.
   a. Estimate the solution of the equation.
   b. What does this solution represent?

30. Nicki spends approximately $4.15 on lunch each day. After a few weeks, she had spent $55.95 on lunches. Let d represent the number of days Nicki bought lunch.
   a. Write an equation using the information given.
   b. How many days did Nicki buy lunch? Show all work necessary to justify your answer.

31. The largest parade of tow trucks on record occurred in Wenatchee, Washington in 2004. Eighty-three tow trucks paraded through the streets. If California has 52 tow trucks committed to being part of a California Tow Truck Parade, how many more tow trucks need to be gathered to tie the Washington record? Explain how you know your answer is correct.

REVIEW

Copy and complete each table. Determine if each number in the table is a solution of the equation.

32.

x	$x + 21 = 30$	Solution?
6		
8		
9		

33.

x	$9x = 45$	Solution?
4		
5		
6		

34. Create three different expressions using x, y and/or z that have a value of 15 if $x = 1$, $y = 2$ and $z = 7$. Use mathematics to prove your expressions have a value of 15.

TIC-TAC-TOE ~ GOING TO THE COUNTY FAIR

Fifteen people in Nate's extended family went to the County Fair. The cost of admission was $5.50 for children and $8 for adults. The total cost of the outing was $90. Based on this information, Nate wants to determine how many children and how many adults from his family attended the County Fair.

Determine how many tickets of each type were purchased by Nate's family. Show all work. Write your final answer in a complete sentence.

SOLVING ADDITION EQUATIONS

LESSON 3.3

Solve equations involving addition.

To solve an equation the variable must be isolated on one side of the equation. This process is sometimes referred to as "getting the variable by itself". The most important thing to remember is that the equation must always remain balanced. Whatever operation is performed on one side of the equation MUST be performed on the other side to keep both sides equal.

THE SUBTRACTION PROPERTY OF EQUALITY

For any numbers a, b and c:
If $a = b$, then $a - c = b - c$

To illustrate the Subtraction Property of Equality, look at this simple equation.

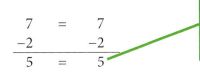

$$\begin{array}{ccc} 7 & = & 7 \\ -2 & & -2 \\ \hline 5 & = & 5 \end{array}$$

Subtract the same number from both sides of the equals sign and the equation remains true.

EXPLORE!

INTRODUCTION TO EQUATION MATS

Step 1: If you do not have an equation mat, draw one like the one seen below on a blank sheet of paper.

=

Step 2: On your equation mat, place a variable cube on one side with 4 chips. On the other side of the mat place 9 chips. This represents the equation $x + 4 = 9$.

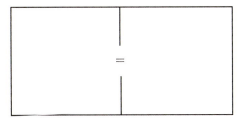

Step 3: To get the variable by itself you must remove the 4 chips that are with the variable cube. If you take chips away from one side of the mat, you must do the same on the other side of the mat (Subtraction Property of Equality). How many chips remain on the right side once 4 have been removed from each side? What does this represent?

Step 4: Clear your mat and place chips and a cube on the mat to represent the equation $x + 3 = 5$. Draw this on paper.

Step 5: What must you do to get the variable cube by itself? Remember that whatever you do on one side of the mat must be done on the other side. How many chips does the variable cube equal? Write your answer in the form $x = $ _____.

Step 6: Create an equation on the mat. Record the algebraic equation on your paper.

Step 7: Solve your equation. What does your variable equal?

Step 8: Write a few sentences to describe how to solve a one-step addition equation using the equation mat, chips and variable cubes.

You will not always have chips, variable cubes and an equation mat available to solve equations. You can solve equations using **inverse operations**. Inverse operations are operations that undo each other. The inverse operation of addition is subtraction.

If an equation has a number being added to the variable, you must subtract the number from both sides of the equation to perform the inverse operation. It is important that you show your work as problems get more complex as you get into higher-level mathematics.

EXAMPLE 1

Solve each equation. Show your work and check your solution.

a. $x + 4 = 11$ **b.** $23 + m = 39$ **c.** $9.4 = y + 5.1$

SOLUTIONS

Draw a vertical line through the equals sign to help you stay organized. Any operation done on one side of the line to cancel out a value must be done on the other side of the line.

> Notice that the variable can be on either side of the equals sign.

a.
$$\begin{array}{r} x + 4 = 11 \\ \underline{-4 \quad -4} \\ x = 7 \end{array}$$

b.
$$\begin{array}{r} 23 + m = 39 \\ \underline{-23 \quad -23} \\ m = 16 \end{array}$$

c.
$$\begin{array}{r} 9.4 = y + 5.1 \\ \underline{-5.1 \quad -5.1} \\ 4.3 = y \end{array}$$

Check your answer by substituting your solution into the original equation for the variable.

☑ $7 + 4 \overset{?}{=} 11$
  $11 = 11$

☑ $23 + 16 \overset{?}{=} 39$
  $39 = 39$

☑ $9.4 \overset{?}{=} 4.3 + 5.1$
  $9.4 = 9.4$

EXERCISES

Solve each equation using an equation mat or inverse operations. Show all work necessary to justify your answer.

1. $x + 6 = 13$

2. $y + 18 = 30$

3. $19 - p = 11$

4. $h + 3.4 = 7.6$

5. $k + 8 = 72$

6. $x + \frac{2}{3} = 1$

7. $427 + y = 533$

8. $4 = d + 4$

9. $f + 0.8 = 2$

10. $19 = m + 1$

11. $p + \frac{1}{4} = 2\frac{3}{4}$

12. $x + 103 = 156$

13. Kyle **did not** check his solutions for the four-problem quiz on one-step equations. Check Kyle's answers. If the answer is incorrect, find the correct answer.

    a. $x + 82 = 124$
       Kyle's answer: $x = 206$

    b. $6 = x + 6$
       Kyle's answer: $x = 1$

    c. $x + 2.6 = 4.9$
       Kyle's answer: $x = 2.3$

    d. $x + \frac{1}{2} = 3\frac{1}{2}$
       Kyle's answer: $x = 3$

14. Tonya is thinking of two numbers. The sum is 34. One of the numbers is 19. What is the other number? Write an equation and solve using inverse operations.

Write an algebraic equation for each word phrase. Solve each equation using an equation mat or inverse operations.

15. The sum of x and twelve equals twenty-six.

16. Seventeen more than x is twenty.

17. Eight plus x equals thirty-one.

18. The sum of x and one-half is five.

19. Dan rents an apartment for $215 more than Jared pays for his rent. Dan's rent is $800.

    a. Explain how the equation $j + 215 = 800$ relates to this situation. What does the variable represent?
    b. How much is Jared's rent?

20. Jillian puts a few apples in a bag and weighs the bag on a produce scale at the grocery store. The bag full of apples weighs 2.25 pounds. She adds one more apple to the bag. The new weight of the bag is 2.6 pounds. Write and solve an addition equation to find the weight of the last apple added to the bag.

21. The distance from Missoula, Montana to Mt. Rushmore is 105 miles more than the distance from St. Paul, Minnesota to Mt. Rushmore. The distance from Missoula to Mt. Rushmore is 730 miles. Write and solve an addition equation to find the distance from St. Paul, Minnesota to Mt. Rushmore.

22. Explain in words how the Subtraction Property of Equality is used to solve an addition equation like $x + 7 = 16$.

Use the order of operations to evaluate.

23. $4 + 5(6) \div 2$

24. $(3 + 7)^2 - 25$

25. $20 - 3 \cdot 2 + 1$

26. $5(5 - 2) + 10 \div 5$

27. $\dfrac{4 + 23}{6 - 3}$

28. $\dfrac{5 + 5^2}{3} + 4$

Use the formula $d = rt$ to evaluate distances.

29. A snail travels at a speed of 0.03 miles per hour. How far will the snail travel in five hours?

30. Kyeron ran 7 miles per hour for half of an hour. How far did he run?

TIC-TAC-TOE ~ DESIGN A WORKOUT

Dee and Molly want to burn 300 calories per day by doing at least one of the exercises described below. Molly likes to do just one activity each day. Dee wants to design workouts that include at least two different activities to reach her calorie goal. Assume Dee and Molly each weigh 100 pounds.

1. Determine how many minutes Molly would need to exercise to burn 300 calories for each activity.

2. Design five different workouts for Dee that include at least two different activities. Each set of activities should help Dee burn approximately 300 calories. Specify how many minutes she will need to complete of each exercise.

Exercise	Calories Burned in One Minute for a 100 pound person
Hacky Sack	3.2
Running (9 minutes per mile)	8.7
Bicycling (12-14 miles per hour)	6.4
Weight Lifting	4.8
Walking (15 minutes per mile)	4.0
Swimming Laps	5.6
Racquetball	7.9

Source: http://www.bodybuilding.com

SOLVING SUBTRACTION EQUATIONS

Solve equations involving subtraction.

Addition and subtraction are inverse operations. In **Lesson 3.3**, you were shown how addition equations can be solved by subtracting a number from both sides of the equation to get the variable by itself. In the same way, subtraction equations can be solved by adding a number to both sides of the equation to get the variable by itself.

Sam's age is $3\frac{1}{2}$ years less than his sister's age. If Sam is 5 years old, how old is his sister? If x represent the sister's age, the equation that represents this situation is

$$x - 3\frac{1}{2} = 5.$$

This equation can be solved using the Addition Property of Equality.

$$\begin{array}{rcl} x - 3\frac{1}{2} &=& 5 \\ +3\frac{1}{2} && + 3\frac{1}{2} \\ \hline x &=& 8\frac{1}{2} \end{array}$$

THE ADDITION PROPERTY OF EQUALITY

For any numbers a, b and c:
If $a = b$, then $a + c = b + c$

EXAMPLE 1

Solve each equation. Check your solution.

a. $y - 3 = 14$ **b.** $52 = x - 7$ **c.** $m - 6\frac{1}{2} = 2$

SOLUTIONS

Draw a vertical line through the equals sign as a reminder that whatever is done on one side of the line must be done on the other side of the line also.

a. $\begin{array}{rcl} y - 3 &=& 14 \\ +3 && +3 \\ \hline y &=& 17 \end{array}$

b. $\begin{array}{rcl} 52 &=& x - 7 \\ +7 && +7 \\ \hline 59 &=& x \end{array}$

c. $\begin{array}{rcl} m - 6\frac{1}{2} &=& 2 \\ +6\frac{1}{2} && + 6\frac{1}{2} \\ \hline m &=& 8\frac{1}{2} \end{array}$

Check the answer by substituting the solution into the original equation for the variable.

☑ $17 - 3 \overset{?}{=} 14$
$14 = 14$

☑ $52 \overset{?}{=} 59 - 7$
$52 = 52$

☑ $8\frac{1}{2} - 6\frac{1}{2} \overset{?}{=} 2$
$2 = 2$

EXERCISES

Solve each equation using inverse operations. Show all work necessary to justify your answer.

1. $y - 9 = 12$

2. $x - 17 = 3$

3. $15 = t - 23$

4. $m - 4.5 = 6.9$

5. $k - 1 = 39$

6. $p - \frac{1}{4} = \frac{1}{2}$

7. $238 = d - 340$

8. $x - 2.05 = 3.4$

9. $f - 68 = 3$

10. $0.5 = b - 2.4$

11. $f - 2\frac{2}{3} = 3\frac{1}{3}$

12. $10 = y - 90$

13. Carrie spent $32 at the movie theater. She had $15 left in her pocket.
  a. If y represents the original amount of money Carrie took to the movie theater, explain why the equation $y - 32 = 15$ represents this situation.
  b. Solve this equation using inverse operations.
  c. What does the answer to this equation represent?

14. Each student has been asked to sell candy bars for the school fundraiser. Bill needs to sell 12 more candy bars to reach his goal. He has already sold 33 candy bars.
  a. Let c represent the number of candy bars Bill wants to sell. Write a subtraction equation for this situation.
  b. Solve the equation. How many candy bars did he want to sell?

15. Terrence was helping give turkeys to needy families at Thanksgiving. Terrence gave away nine turkeys. There were sixteen turkeys remaining.
  a. Let t represent the number of turkeys Terrence started with. Write a subtraction equation for this situation.
  b. Solve the equation. How many turkeys did Terrence have at the beginning?
  c. Terrence gave away all the turkeys in five hours. Approximately how many turkeys did he give per hour? Show all work necessary to justify your answer.

Write an algebraic equation for each sentence. Solve each equation.

16. A number x minus six equals twenty-one.

17. Seventy-two is ten less than x.

18. Nineteen is five less than x.

19. A number x minus three-fourths is one-half.

20. Pioneers traveled across the Oregon Trail to settle in new parts of the United States during the 19th century. After the pioneers traveled 1,240 miles they still had 930 miles left to go to get to Oregon.
  a. Let x represent the total distance the pioneers traveled on the Oregon Trail. Write a subtraction equation for the situation.
  b. Solve the equation. How many total miles did the pioneers travel across the Oregon Trail?

21. Create a situation that can be solved using a subtraction equation. Explain the situation and state the answer to your equation.

22. Four equations are listed below. Determine which equations you can solve with the Addition Property of Equality and which equations require the Subtraction Property of Equality. Solve all four equations.

    a. $x + 42 = 89$ **b.** $y - 15 = 6$ **c.** $56 = p - 8$ **d.** $m + 0.3 = 7$

REVIEW

Use the following formulas to answer the questions about the drawing below.

Area = $\frac{1}{2}bh$

Area = lw
Perimeter = $2l + 2w$

Area = πr^2
Circumference = $2\pi r$

23. Find the area of the triangular roof at the front of the house.

24. Find the circumference of one of the circular windows given that the radius of each window is 2 feet.

25. Find the area of each circular window given that the radius of each window is 2 feet.

26. Find the perimeter of the rectangular front panel of the house.

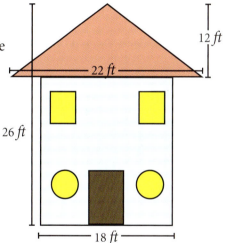

12 ft

22 ft

26 ft

18 ft

TIC-TAC-TOE ~ FLIP BOOK

Solving One-Step Equations

Pretend that a new student will enroll in your school tomorrow and will need to be taught how to solve all types of one-step equations. Create a flip book that teaches how to solve the four types of one-step equations. Make sure the flip book helps students know which operation to choose. Include examples.

Solve equations involving multiplication or division.

A balance scale is another way to represent how two sides of an equation are equal in amount. Solving equations is like keeping a scale balanced. Imagine that the equals sign is the balancing point on the scale. Any operation performed on one side of the scale must also be performed on the other side to keep the scale balanced.

So far in this block you have learned how to solve addition and subtraction equations. In this lesson, you will work with multiplication and division equations. Multiplication and division are inverse operations. To solve a multiplication equation, you will divide each side of the equation by a number to get the variable by itself. To solve a division equation, you will use multiplication.

EXPLORE!

MULTIPLICATION EQUATIONS

Step 1: If you do not have an equation mat, draw one like the one seen on the right on a blank sheet of paper.

Step 2: On the equation mat, place two variable cubes on one side. On the other side of the mat, place 6 chips. This represents the equation $2x = 6$.

Step 3: Divide the chips into two equal groups since there are two variable cubes. How many chips are equal to one variable cube? This is the value of x.

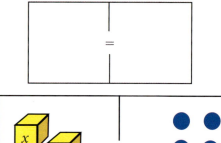

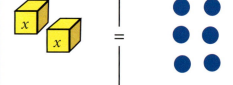

Step 4: Clear the mat and place counters on the mat to represent the equation $4x = 8$. Draw this on your paper.

Step 5: How can you determine how many chips are equal to one variable cube?

Step 6: Create a multiplication equation on the mat. Record the algebraic equation on your paper.

Step 7: Solve the equation. What does the variable equal?

Step 8: Write a few sentences that describe how to solve a multiplication equation using the equation mat, chips and variable cubes.

THE MULTIPLICATION PROPERTY OF EQUALITY

For any numbers a, b and c:
If $a = b$, then $a \cdot c = b \cdot c$.

THE DIVISION PROPERTY OF EQUALITY

For any numbers a, b and c:
If $a = b$, then $\dfrac{a}{c} = \dfrac{b}{c}$.

EXAMPLE 1

Solve each equation. Show your work and check your solution.

a. $\dfrac{m}{8} = 3$　　　　　　　　　b. $6.2 = \dfrac{y}{3}$

c. $11p = 99$　　　　　　　　　　d. $12x = 60$

SOLUTIONS

The fraction bar is used to show division when solving equations.

a. $\dfrac{m}{8} = 3$

$8 \cdot \dfrac{m}{8} = 3 \cdot 8$

$m = 24$

$\boxed{\checkmark}\ \dfrac{24}{8} = 3$

b. $6.2 = \dfrac{y}{3}$

$3 \cdot 6.2 = \dfrac{y}{3} \cdot 3$

$18.6 = y$

$\boxed{\checkmark}\ 6.2 = \dfrac{18.6}{3}$

c. $\dfrac{11p}{11} = \dfrac{99}{11}$

$p = 9$

$\boxed{\checkmark}\ 11(9) = 99$

d. $\dfrac{12x}{12} = \dfrac{60}{12}$

$x = 5$

$\boxed{\checkmark}\ 12(5) = 60$

EXAMPLE 2

Peter is four times older than his son, Matthew. Peter is 52 years old. Write a multiplication equation and solve the equation to determine Matthew's age.

SOLUTION

Let m represent Matthew's age. Write the equation that represents this situation.

$$4m = 52$$

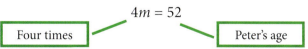

Four times　　　　Peter's age

Solve the equation by using division (the inverse operation of multiplication).

$\dfrac{4m}{4} = \dfrac{52}{4}$

$m = 13$

Matthew is 13 years old.

EXERCISES

Determine which operation (multiplication or division) would be used to solve each equation.

1. $\frac{h}{7} = 12$

2. $3x = 57$

3. $14 = 4m$

Solve each equation using inverse operations. Show all work necessary to justify your answer.

4. $6y = 12$

5. $2p = 24$

6. $\frac{m}{3} = 7$

7. $\frac{h}{10} = 8$

8. $42 = 7x$

9. $5p = 35$

10. $14j = 28$

11. $4 = \frac{b}{11}$

12. $\frac{y}{6} = 0.5$

13. $2.5m = 10$

14. $\frac{c}{20} = 8$

15. $0.9x = 7.2$

Write an algebraic equation for each sentence. Solve each equation.

16. Six times a number y is sixty-six.

17. A number x divided by four is eight.

18. The quotient of x divided by seven is five.

19. Thirty-six is nine times a number x.

20. Oscar is three times older than James. Oscar is 42 years old.
　　a. Let j represent James' age. Write a multiplication equation for this situation.
　　b. Solve the equation. How old is James?

21. Hannah is five times as old as her sister. Hannah is 15 years old. Write a multiplication equation. Solve the equation to determine her sister's age.

22. Three friends share a pizza. Each friend eats 4 pieces.
　　a. Let p represent the total number of pieces of pizza. Write a division equation for this situation.
　　b. Solve the equation. How many pieces of pizza were there in all?

23. Five people purchased a piece of farm land together. Ten years later they sold the land. Each person received $45,217. What was the total amount the land was sold for? Explain how you know your answer is correct.

24. Dora weighs $5\frac{1}{2}$ times as much as her baby brother, Jeremie. Dora weighs $60\frac{1}{2}$ pounds. How much does Jeremie weigh? Show all work necessary to justify your answer.

Match each expression with an equivalent expression.

25. $2(x - 7)$ **A.** $7x + 6$

26. $4x + 3 + 8x + 1$ **B.** $12x + 4$

27. $7(x + 1) - 1$ **C.** $7x + 8$

28. $2(x + 3) + 5x + 2$ **D.** $2x - 14$

TIC-TAC-TOE ~ MISSING PRICE TAGS

Kiera went to the store and purchased six items. All items were on sale but the original price tags had been removed. Use the information shown below to help Kiera determine the original price of each item she purchased. Use the given percent equation. Percents must be converted to decimal form.

original price · percent savings = amount saved

1. Determine the original cost of each item.
2. Determine the total cost for all six items before they went on sale.
3. Determine the total amount Kiera spent on the sale items.

20% off Save $1.60

40% off Save $7.20

25% off Save $4.00

15% off Save $4.50

60% off Save $5.40

30% off Save $6.90

TIC-TAC-TOE ~ LIKE TERMS

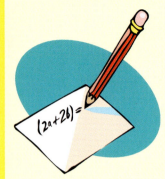

Some equations have like terms that must be combined before you can use inverse operations to isolate the variable. There are two types of equations you may deal with that have like terms:
1) Equations that have like terms on the same side of the equals sign.
2) Equations that have like terms on opposite sides of the equals sign.

Type 1: When equations have like terms on the same side of the equals sign, you must first combine the like terms before using inverse operations to isolate the variable.

For example:

$5x + 2x - 3x = 20$

$5x + 2x - 3x$ combines to be $4x$.

$4x = 20$

$\dfrac{4x}{4} = \dfrac{20}{4}$

$x = 5$

Type 2: Equations with like terms on each side of the equals sign require you to move the variables to one side of the equation and the constants to the other side of the equation. Whenever a term is moved from one side of the equals sign to the other, the inverse operation must be used.

Step 1: Move the variable term with the smaller coefficient to the other side of the equals sign using inverse operations.
Step 2: Move all constants away from the variable.
Step 3: Use division to solve.

For example:

$$
\begin{array}{rcl}
9x - 5 &=& 6x + 7 \\
-6x & & -6x \\
\hline
3x - 5 &=& 7 \\
+5 & & +5 \\
\hline
\dfrac{3x}{3} &=& \dfrac{12}{3} \\
x &=& 4
\end{array}
$$

Solve each equation. Show all work necessary to justify your answer.

1. $8x - 3x + 4x = 72$

2. $2x + 22 = 6x + 2$

3. $4 + 11 + 15 = 5x$

4. $12 + 3x = x + 42$

5. $10x - 41 = 2x + 7$

6. $27 = 8x + 2x - 7x$

7. $3x + 18 = 9x - 3$

8. $x + 2x + 3x = 4 + 7 + 1$

9. $2x = 8x - 24$

MIXED ONE-STEP EQUATIONS

LESSON 3.6

Solve a variety of one-step equations.

EXPLORE!   **INVERSE OPERATIONS**

You have now learned how to solve four different types of equations (addition, subtraction, multiplication and division). The key to solving equations is to determine which operation to use to get the variable by itself.

Step 1: Fold a piece of paper in half vertically. Open up the paper and fold it in half horizontally. Open the paper up and lay it flat on a desk.

Addition Equations	Subtraction Equations
Multiplication Equations	Division Equations

Step 2: Copy the diagram at the right. At the top of one section, write the words ADDITION EQUATIONS. At the top of another section, write the words SUBTRACTION EQUATIONS, and so on.

Step 3: Twelve equations are below. Classify each equation by type and write it in the appropriate box on the paper. Leave space to solve the equations.

$x - 7 = 24$ $\dfrac{y}{4} = 13$ $p + 19 = 21$

$16 = m + 5$ $3k = 36$ $\dfrac{w}{22} = 5$

$100 = 25b$ $u - \dfrac{1}{5} = \dfrac{3}{10}$ $39 + g = 100$

$3 = \dfrac{z}{5}$ $3.1c = 12.4$ $a - 210 = 134$

Step 4: Solve each equation and show your work.

Step 5: Write the inverse operation used to solve each type of equation at the bottom of each box.

Step 6: Explain to a classmate how you know which operation to choose when solving an equation.

EXAMPLE 1

Write an algebraic equation for each sentence. Solve each equation.
a. The sum of a number x and seven is thirty-two.
b. A number y divided by four is thirteen.
c. Ten less than a number b is twenty-five.
d. The product of a number p and eight is twenty.

SOLUTIONS

a. Sum is another word for addition.
Write the equation.
Solve. Subtraction is the inverse operation of addition.

$$x + 7 = 32$$
$$\underline{-7 \mid -7}$$
$$x = 25$$

b. Write the equation.

$$\frac{y}{4} = 13$$

Solve. Multiplication is the inverse operation of division.

$$4 \cdot \frac{y}{4} = 13 \cdot 4$$
$$y = 52$$

c. Write the equation.
Solve. Addition is the inverse operation of subtraction.

$$b - 10 = 25$$
$$\underline{+10 \mid +10}$$
$$b = 35$$

d. Product is another word for multiplication.
Write the equation.
Solve. Division is the inverse operation of multiplication.

$$8p = 20$$
$$\frac{8p}{8} = \frac{20}{8}$$
$$p = 2.5$$

For application problems, also called "story problems," write an equation to help solve the problem. It is important to locate key words in the problem to determine which operation to use.

Addition and Subtraction
Are items being compared with words like "more than" or "less than"? Are words like "sum", "difference" or "increased or decreased by" being used? You might need to write an addition or subtraction equation.

Multiplication and Division
Are you dividing people or items or putting them into groups? Do you see words like "times", "product" or "quotient"? You might need to write a multiplication or division equation.

EXERCISES

Determine which operation (addition, subtraction, multiplication or division) to use to solve each equation.

1. $15m = 165$

2. $122 = w - 58$

3. $x + 5.6 = 10.2$

4. $\dfrac{b}{12} = 2$

5. $7 + k = 15$

6. $6 = 4y$

Solve each equation using inverse operations. Check the solution.

7. $x + 12 = 43$

8. $40 = 5y$

9. $p - 3.7 = 4.2$

10. $32f = 64$

11. $3 = \dfrac{k}{9}$

12. $y + 1\frac{1}{2} = 3\frac{3}{4}$

13. $m + 429 = 530$

14. $4 = c - 91$

15. $\dfrac{x}{3} = 30$

16. $4w = 20$

17. $y - 5.04 = 6.2$

18. $\dfrac{d}{8} = 0.25$

19. $5 + a = 5$

20. $\dfrac{b}{6} = 3$

21. $0.9x = 7.2$

22. $\dfrac{h}{7} = 101$

23. $0 = y - 14$

24. $8p = 8$

Write an algebraic equation for each sentence. Solve each equation.

25. The sum of a number y and eight is twenty.

26. A number x divided by six is eleven.

27. The product of x and fifteen equals 45.

28. A number p decreased by fourteen is three.

29. The quotient of x divided by two is fifty.

30. Twenty increased by a number x is 44.

31. Valencia's hourly wage is 3.2 times the amount of Horatio's hourly wage. Valencia makes $31.20 per hour.
   a. Let h represent Horatio's wage. Write an equation for this situation.
   b. Solve the equation. What is Horatio's hourly wage?

32. The circus was almost sold out. They had sold 4,030 tickets. There were 570 unsold tickets left.
   a. Let y represent the total number of tickets that could be sold. Write an equation for this situation.
   b. How many people will be at the circus if all tickets are sold? Explain how you know your answer is correct.

33. It is spirit week at McKinley Middle School. The seventh graders were in the lead until they lost 16 points on Wednesday for poor sportsmanship at the pep assembly. After 16 points were taken away, the seventh grade team still had 19 points.

    a. Let p represent the total number of points the seventh grade team had before the pep assembly. Write an equation for this situation.

    b. Solve the equation. How many points did the seventh grade team have before the assembly?

34. The average new house provides 3 times the amount of square footage per person than houses did in 1950. The average new house currently provides approximately 870 square feet per family member. What was the average square footage per person in 1950? Show all work necessary to justify your answer.

REVIEW

Simplify each algebraic expression by combining like terms. Use the Distributive Property, if necessary.

35. $6x + 4x + x$ **36.** $3 + 2y + 2y + 1$ **37.** $6(y + 3) - 18$

Use the formula $B = \frac{h}{a}$ to evaluate batting averages where B represents the batting average, h is the number of hits and a is the 'at bats'. Round to the nearest thousandth.

38. Find the batting average when $h = 12$ and $a = 50$.

39. The Bakersville Blue Sox team members have been up to bat 478 times this season. The team has 105 hits. What is the team batting average?

TIC-TAC-TOE ~ EQUATION BINGO

Type a worksheet made up of 24 one-step equations. Allow room for solving the equations on the worksheet. Type the answers to the equations on a separate page. None of the answers can be the same.

Create a blank BINGO card that is 5 squares across and 5 squares down with a free space in the center of the card.

To play the game, participants will solve the equations and place their answers anywhere on the BINGO card. The person in charge of the activity will call out the answers to equations in no particular order. Participants cross off the boxes with the corresponding answers. The game can be played for a normal BINGO. If desired, continue playing for a blackout. Each participant who solved all 24 equations correctly will get a blackout.

FORMULAS AND EQUATION-SOLVING

 Solve formulas for a variety of situations.

Gregory knows his batting average for the season was 0.314. His coach told him he had 70 'at bats' during the season but was unsure of how many hits he had. Gregory figured it out himself:

He wrote out the batting average formula.	$B = \dfrac{h}{a}$
He substituted his batting average for B and his 'at bats' for *a*.	$0.314 = \dfrac{h}{70}$
He solved the formula.	$70 \cdot 0.314 = \dfrac{h}{\cancel{70}} \cdot \cancel{70}$
	$21.98 = h$

Gregory rounded to the nearest whole number. $\qquad\qquad 21.98 \approx 22$

Gregory had 22 hits this season.

Formulas are very useful in calculating quantities such as area, perimeter, interest earned in a bank account, distance and batting average. In **Block 2** you evaluated these formulas when you were given values to substitute for the variables. In this lesson you will be able to work backwards to find a missing value in the equation by using your equation-solving skills.

WORKING WITH EQUATIONS TO SOLVE FOR A MISSING VALUE

1. Find the formula that fits the situation given.
2. Substitute all known values for the variables into the formula.
3. Determine if any numbers can be multiplied that are on the same side of the equals sign. If so, multiply those numbers.
4. Solve the equation using inverse operations.
5. Write out the answer in a complete sentence.

Geometry formulas to use in this lesson:

Area = $\frac{1}{2}bh$

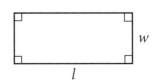

Area = *lw*

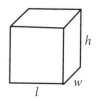

Volume = *lwh*

EXAMPLE 1

The area of a triangle is 36 square units. The base is 8 units long. What is the height of the triangle?

SOLUTION

Write the formula.

$$\text{Area} = \tfrac{1}{2}bh$$

Substitute values for the variables.

$$\text{Area} = \tfrac{1}{2}bh$$

$$36 = \tfrac{1}{2}(8)h$$

Two numbers need to be multiplied on the right-hand side of the equation before solving.

$$36 = 4h$$

Solve the equation using inverse operations.

$$\frac{36}{4} = \frac{4h}{4}$$

$$9 = h$$

The height of the triangle is 9 units.

More formulas:

$$I = prt$$

I = interest
p = initial deposit
r = rate (as a decimal)
t = time

$$d = rt$$

d = distance
r = rate
t = time

$$B = \frac{h}{a}$$

B = batting average
h = hits
a = 'at bats'

EXAMPLE 2

Kenny deposited $500 in a bank account. After 3 years, Kenny earned $90 interest. Use the simple interest formula to determine the interest rate for this account.

SOLUTION

Write the formula.

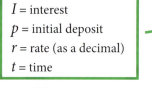

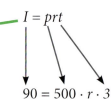

$$I = prt$$

Substitute values for the variables.

$$90 = 500 \cdot r \cdot 3$$

Two numbers can be multiplied on the right-hand side of the equation. Multiply before solving.

$$90 = 500 \cdot r \cdot 3$$
$$90 = 1500 \cdot r$$

Solve the equation using inverse operations.

$$\frac{90}{1500} = \frac{1500r}{1500}$$

$$0.06 = r$$

Change the decimal to a percent by multiplying by 100.

$$0.06(100) = 6\%$$

Kenny earned an interest rate of 6%.

EXAMPLE 3

The volume of a rectangular box is 60 cubic inches. The length of the box is 5 inches and the height of the box is 2 inches. Find the width of the box.

SOLUTION

Write the formula.	Volume = lwh
Substitute values for the variables.	$60 = (5)w(2)$
Multiply the two numbers on the right-hand side before solving.	$60 = (5)w(2)$ $60 = 10w$
Solve the equation using inverse operations.	$\dfrac{60}{10} = \dfrac{\cancel{10}w}{\cancel{10}}$ $6 = w$

The box has a width of 6 inches.

EXERCISES

Formulas to use in the exercises:

Area = $\frac{1}{2}bh$

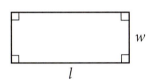

Area = lw

Volume = lwh

$I = prt$

$d = rt$

$B = \dfrac{h}{a}$

Find the missing value for each equation. Show all work necessary to justify your answer. State each answer in a complete sentence.

1. The area of a rectangle is 45 square units. The width of the rectangle is 9 units. Find the length.

2. Ryan had 25 'at bats' this season. His batting average is 0.400. Find how many hits he had this season.

3. The volume of a box of Jenny's favorite cereal is 378 cubic inches. The height of the box is 14 inches. The length of the box is 9 inches. Find the width of the box.

4. Jakin traveled a distance of 175 miles in 5 hours. At what rate was he driving?

5. The height of a triangle is 12 centimeters. The area of the triangle is 48 square centimeters. Find the length of the base of the triangle.

6. Maria put $100 in a savings account that earns simple interest. Her account earns 4% interest. She withdrew the money after she had earned $16 in interest. How many years did she leave her money in the account?

7. While Emma was on vacation, she walked on the beach for 2 hours. She covered a total distance of 5 miles. At what rate was Emma walking?

8. Toby measured the length of his rectangular yard and found that it was 8 meters long. He knows the area of his yard is 44 square meters. What is the width of his yard?

9. Lisa invested $800 for 5 years in a bank account that earns simple interest. She earned $140 in interest. What interest rate was she getting in this account?

10. Owen calculated the volume of a rectangular prism to be 120 cubic units. He lost the measurement for the height of the prism. Determine the height of the prism given the information on the figure at right.

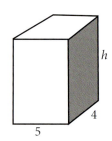

11. Luke went on a sailboat with his family on the Pacific Ocean. The base of the triangular sail measured 14 feet. The area of the sail was 224 square feet. Find the height of the sail.

12. A rectangular prism has a height of 4 *cm*. Its volume is 48 *cm³*. Find three different pairs of possible dimensions for the length and width of the prism if all dimensions have whole number values. Explain how you know your answers are correct.

13. The area of a triangle is 24 square feet. The height of the triangle is 6 feet. Keira determined the base is 2 feet using the math below. What did she do wrong? What is the length of the base?

$$24 = \tfrac{1}{2}bh$$
$$24 = \tfrac{1}{2}b(6)$$
$$4 = \tfrac{1}{2}b$$
$$2 = \text{base}$$

REVIEW

Solve each equation using inverse operations. Show all work necessary to justify your answer.

14. $y - 115 = 228$

15. $4x = 44$

16. $k + 7.2 = 12.2$

17. $72 = 2m$

18. $2.7 = \frac{p}{10}$

19. $80 = c - 24$

20. $y + 4 = 53$

21. $d + 2\tfrac{1}{8} = 3\tfrac{1}{2}$

22. $\frac{x}{5} = 8$

23. $15w = 75$

24. $y - 0.2 = 0.45$

25. $\frac{d}{8} = 1.5$

SOLVING TWO-STEP EQUATIONS

 Solve two-step equations.

Julie loves to attend the Spring Carnival every year in her hometown. Julie paid $9 to enter the carnival. Each ride cost her $4. She spent a total of $29. How many rides did she go on?

Let x represent the number of rides that Julie went on. The equation that models this situation is:

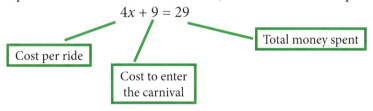

$$4x + 9 = 29$$

Cost per ride

Cost to enter the carnival

Total money spent

This equation is called a **two-step equation**. Two-step equations have two different operations in the equation. The two-step equation above has both multiplication and addition. In order to get the variable by itself you must perform two inverse operations. First you must balance the equation by undoing any ADDITION or SUBTRACTION. You can finish solving the equation by undoing any MULTIPLICATION or DIVISION. It is done in this order because it is undoing the order of operations.

EXPLORE!   EQUATION MATS FOR TWO-STEP EQUATIONS

Step 1: If you do not have an equation mat, draw one like the one seen at right on a blank sheet of paper.

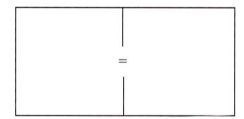

Step 2: On your equation mat place two variable cubes on one side with 4 chips. On the other side of the mat place 6 chips. What two-step equation does this represent?

Step 3: The first step to solving an equation is to remove the chips on the side with the variable. Remove an equal number of chips from BOTH sides of the mat. Draw a picture of what your mat looks like now.

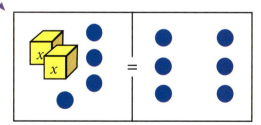

Step 4: Divide the chips that are left on the mat equally between the two variable cubes. How many chips are equal to one cube? This is the value of x.

Step 5: Solve the equation $3x + 2 = 11$ using the equation mat. What does x equal?

EXAMPLE 1

Solve each equation. Check the solution.

a. $7x - 3 = 74$ **b.** $\dfrac{x}{9} + 5 = 7$

SOLUTIONS

a.
$$7x - 3 = 74$$
$$\underline{+3 \mid +3}$$
$$\frac{7x}{7} = \frac{77}{7}$$

$$x = 11$$

☑ $7(11) - 3 \overset{?}{=} 74$
$$77 - 3 \overset{?}{=} 74$$
$$74 = 74$$

b.
$$\frac{x}{9} + 5 = 7$$
$$\underline{\phantom{\frac{x}{9}}-5 \mid -5}$$
$$9 \cdot \frac{x}{9} = 2 \cdot 9$$

$$x = 18$$

☑ $\dfrac{18}{9} + 5 \overset{?}{=} 7$
$$2 + 5 \overset{?}{=} 7$$
$$7 = 7$$

EXAMPLE 2

Write an equation for each statement. Solve each problem and check the solution.
a. Three times a number x increased by 8 is 29. Find the number x.
b. Twice a number x minus 7 is 5. Find the number x.

SOLUTIONS

a. Three times a number $x \rightarrow 3x$
increased by $8 \rightarrow 3x + 8$
is $29 \rightarrow 3x + 8 = 29$
$$3x + 8 = 29$$
$$\underline{-8 \mid -8}$$
$$\frac{3x}{3} = \frac{21}{3}$$

$$x = 7$$

☑ $3(7) + 8 \overset{?}{=} 29$
$$21 + 8 \overset{?}{=} 29$$
$$29 = 29$$

b. Twice a number $x \rightarrow 2x$
minus $7 \rightarrow 2x - 7$
is $5 \rightarrow 2x - 7 = 5$
$$2x - 7 = 5$$
$$\underline{+7 \mid +7}$$
$$\frac{2x}{2} = \frac{12}{2}$$

$$x = 6$$

☑ $2(6) - 7 \overset{?}{=} 5$
$$12 - 7 \overset{?}{=} 5$$
$$5 = 5$$

EXAMPLE 3

John and Erin shared a bag of chocolate candies. John ate five fewer than three times as many candies as Erin ate. If John ate 43 candies, how many candies did Erin eat?

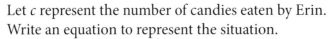

SOLUTION

Let c represent the number of candies eaten by Erin.
Write an equation to represent the situation.

$$3c - 5 = 43$$

Solve the equation using inverse operations.

$$3c - 5 = 43$$
$$\underline{+5 \mid +5}$$
$$\frac{3c}{3} = \frac{48}{3}$$

Erin ate 16 candies.

$$c = 16$$

Solve each equation for the variable. Show all work necessary to justify your answer.

1. $4x + 2 = 18$

2. $\frac{h}{3} + 1 = 5$

3. $\frac{x}{5} - 7 = 3$

4. $5m + 6 = 31$

5. $6y + 5 = 17$

6. $\frac{w}{4} - 8 = 3$

7. $16 = \frac{p}{4} + 10$

8. $12p - 3 = 21$

9. $3x + 9 = 12$

10. $\frac{a}{2} - 8 = 12$

11. $\frac{w}{5} + 1 = 1$

12. $30x - 14 = 106$

13. $40 = 4x - 4$

14. $\frac{y}{4} + 3 = 5$

15. Xavier solved three problems incorrectly. For each problem, describe the error he made and find the correct answer.

a.
$$3x - 6 = 27$$
$$\underline{\quad -6 \quad | \quad -6 \quad}$$
$$\frac{3x}{3} = \frac{21}{3}$$
$$x = 7$$

b.
$$45 = 20x + 15$$
$$\underline{\quad -15 \quad -15 \quad}$$
$$\frac{45}{5} = \frac{5x}{5}$$
$$x = 9$$

c.
$$\frac{x}{8} - 2 = 22$$
$$\underline{\quad +2 \quad | \quad +2 \quad}$$
$$8 \div \frac{x}{8} = \frac{24}{8}$$
$$x = 3$$

Write an equation for each statement. Find the value of the variable. Show all work necessary to justify your answer.

16. Twice a number x plus seven is twenty-five. Find the number.

17. Four more than a number y divided by two is ten. Find the number.

18. Twelve increased by five times a number w is 72. Find the number.

19. Three less than one-half of a number p is 4. Find the number.

20. Scott had three less than twice as many meatballs on his plate than Beth had on her plate. Scott had seven meatballs on his plate. How many meatballs were on Beth's plate? Use words and/or numbers to show how you determined your answer.

21. Latrelle gets paid an initial payment of $15 each time he babysits at the neighborhood get-together. He also gets paid an additional $2 per child. Last time, he was paid $27. Write and solve an equation to determine how many children he babysat that time.

22. Kathy owns her own business selling homemade jam. Each jar of jam weighs 18 ounces. The box she mails the jam in weighs 6 ounces. How many jars of jam are in a package that weighs 150 ounces? Show all work necessary to justify your answer.

REVIEW

Copy and complete each table by evaluating the given expression for the values listed.

23.

x	$4x - 1$
$\frac{1}{2}$	
3	
5	
12	

24.

x	$\frac{5x + 2}{2}$
2	
3	
6	
100	

25.

x	$(x + 1)^2$
1	
2	
5	
9	

TIC-TAC-TOE ~ MISSING DIMENSIONS

The area of each figure below is given. Find the missing dimension. Use mathematics to justify your answer. Access formulas online or on a formula sheet, if necessary.

1. Area = 124 ft^2

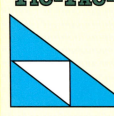

2. Area = 87.25 ft^2

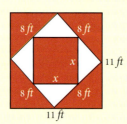

3. Area = 72 cm^2

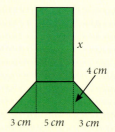

4. Area = 34.4 m^2

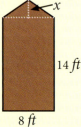

5. Area = 66 m^2

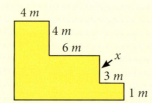

6. Area = 82.8 in^2

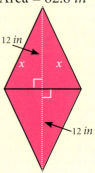

TIC-TAC-TOE ~ EARNING AN 'A' IN MATH

Mr. Blanchard gives five tests each quarter in his math classes. Students must average 90% on the five tests in order to earn an 'A' in his class. Students have taken four of the five tests so far this quarter. Some of the students want to figure out how low a score they can earn on the last test and still earn an 'A'.

You must add all the scores together and divide by the number of tests that were taken to find an average.

For example: Devin scored 82%, 95%, 79% and 96% on his first four tests. He is averaging 88%.

$$\frac{82 + 95 + 79 + 96}{4} = 88$$

To find the lowest score Devin can get on his last test to earn an 'A' in the class he must find the missing test score using the formula to find the average with five tests. Set the formula equal to 90 since Devin is trying to earn an 'A' in the class.

$$\frac{82 + 95 + 79 + 96 + x}{5} = 90$$

Add the numbers in the numerator.

$$\frac{352 + x}{5} = 90$$

Use inverse operations and multiply both sides of the equation by 5.

$$352 + x = 450$$

Use inverse operations and subtract 352 from both sides of the equation.

$$x = 98$$

Devin must score at least a 98% on the last test to earn an 'A'.

Copy and complete the table below. Determine the lowest grade each student must earn on their last test to get an 'A' in Mr. Blanchard's math class. Show all work.

Student	Test #1	Test #2	Test #3	Test #4	Test #5
Larry	100%	92%	88%	96%	?
Misha	83%	98%	90%	85%	?
Sydney	100%	95%	97%	95%	?
Roscoe	82%	85%	89%	94%	?

Vocabulary

equation

inverse operations

solution

two-step equation

Determine if a number is a solution of an equation.
Solve one-step equations using mental math.
Solve equations involving addition.
Solve equations involving subtraction.
Solve equations involving multiplication or division.
Solve a variety of one-step equations.
Solve formulas for a variety of situations.
Solve two-step equations.

Lesson 3.1 ~ Equations and Solutions

Determine if the number given is the solution of the equation.

1. $x - 4 = 3$ *Is 12 the solution?*

2. $4y = 48$ *Is 12 the solution?*

3. $\frac{b}{6} = 5$ *Is 30 the solution?*

4. $h + 26 = 40$ *Is 24 the solution?*

Copy and complete each table. Determine if each number in the table is a solution of the equation.

5.

x	3x = 27	Solution?
7		
8		
9		

6.

x	x + 12 = 23	Solution?
11		
12		
21		

Write an algebraic equation for each sentence. Use *x* as the variable.

7. A number plus eight equals ten.

8. A number minus eleven equals seven.

9. Four times a number is twenty-four.

10. The quotient of a number divided by two is six.

Use Guess, Check and Revise to solve each equation.

11. $2x = 6$

12. $y + 5 = 20$

13. $5x = 35$

14. $18 \div m = 3$

15. $y - 7 = 13$

16. $\frac{w}{10} = 3$

17. Jeremiah bought 4 milkshakes for himself and his friends. He spent a total of $20. Let m represent the price of one milkshake.

    a. Write a multiplication equation using the information given.

    b. Solve the equation using mental math. How much does each milkshake cost?

Lesson 3.3 ~ Solving Addition Equations

Solve each equation using an equation mat or inverse operations. Show all work necessary to justify your answer.

18. $x + 3 = 10$

19. $y + 11 = 35$

20. $8 + m = 9$

21. $a + 2.7 = 3.3$

22. $29 = k + 10$

23. $x + \frac{1}{2} = 2\frac{3}{4}$

Write an algebraic equation for each sentence. Solve each equation.

24. The sum of x and nine equals thirteen.

25. Twenty-two more than x is forty.

26. Kelsey earns $310 more than Michelle each month. Kelsey earns $1250 each month.

    a. Explain how the equation $x + 310 = 1250$ relates to this situation. What does x represent?

    b. How much money does Michelle make each month? Show all work necessary to justify your answer.

Lesson 3.4 ~ Solving Subtraction Equations

Solve each equation using inverse operations. Show all work necessary to justify your answer.

27. $x - 4 = 13$

28. $y - 16 = 12$

29. $1 = p - 45$

30. $m - 129 = 234$

31. $p - \frac{2}{5} = \frac{1}{10}$

32. $k - 4.3 = 3$

Write an algebraic equation for each sentence. Solve each equation.

33. A number *x* minus five is thirty-one.

34. Seventeen less than *x* is forty.

35. Christopher knows he should drink a lot of water to stay healthy. Christopher needs to drink 24 more ounces of water to reach his goal for the day. He has already consumed 42 ounces of water.
    a. Let *w* represent Christopher's goal amount of water in ounces. Write a subtraction equation for this situation.
    b. How many ounces of water does he want to drink each day? Explain how you know your answer is correct.

Lesson 3.5 ~ Solving Multiplication and Division Equations

Solve each equation using inverse operations. Show all work necessary to justify your answer.

36. $7y = 21$

37. $2p = 18$

38. $\frac{m}{5} = 4$

39. $\frac{x}{10} = 7$

40. $4.4 = 1.1b$

41. $\frac{a}{3} = 2.5$

Write an algebraic equation for each sentence. Solve each equation.

42. Four times a number *y* is thirty-six.

43. A number *x* divided by seven is three.

44. Jennifer is three times older than her cousin, Julie. Jennifer is 24 years old.
    a. Let *x* represent Julie's age. Write a multiplication equation for this situation.
    b. Solve the equation. How old is Julie?

45. Marcy ate five times more candy on Halloween than her friend Wesley ate. Marcy ate 15 pieces of candy. How many pieces of candy did Wesley eat on Halloween? Use words and/or numbers to show how you determined your answer.

Lesson 3.6 ~ Mixed One-Step Equations

Determine which operation (addition, subtraction, multiplication or division) would be used to solve each equation.

46. $x + 17 = 128$

47. $12 = y - 82$

48. $15m = 550$

49. $\frac{b}{12} = 15$

50. $9 + w = 15$

51. $8 = 5y$

Solve each equation using inverse operations. Show all work necessary to justify your answer.

52. $x + 7 = 12$

53. $40 = y - 6$

54. $\frac{m}{3} = 12$

55. $22p = 44$

56. $2.3 = \frac{k}{2}$

57. $104 = a - 82$

58. $w + 42 = 53$

59. $y - 1\frac{1}{3} = 3\frac{4}{9}$

60. $\frac{x}{7} = 20$

Lesson 3.7 ~ Formulas and Equation-Solving

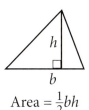

Area = $\frac{1}{2}bh$

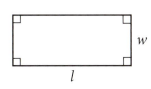

Area = lw

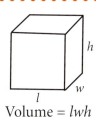

Volume = lwh

$I = prt$

$d = rt$

$B = \frac{h}{a}$

For each exercise, find the missing value and state your answer in a complete sentence.

61. The area of a rectangle is 35 square units. The length of the rectangle is 7 units. Find the width.

62. Frank has 32 'at bats' this season. His batting average is 0.375. Find how many hits he has this season.

63. The volume of a box of Petra's favorite candy is 336 cubic centimeters. The height of the box is 14 centimeters. The length of the box is 8 centimeters. Find the width of the box.

64. Aiden ran 15 miles in 2.5 hours. At what rate was he running?

65. The height of a triangle is 8 centimeters. The area of the triangle is 56 square centimeters. Find the length of the base of the triangle. Explain how you know your answer is correct.

66. Vanessa put $200 in a savings account that earns simple interest. Her account earns 3% interest. She withdrew the money after she earned $24. How many years did she leave her money in the account?

Lesson 3.8 ~ Solving Two-Step Equations

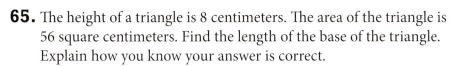

Solve each equation for the variable. Show all work necessary to justify your answer.

67. $5x + 3 = 23$

68. $\frac{y}{2} - 4 = 3$

69. $\frac{a}{6} + 2 = 12$

70. $4m - 7 = 25$

71. $2 = \frac{x}{3} - 10$

72. $\frac{w}{10} + 1 = 3$

Write an equation for each statement. Solve each problem and check your solution.

73. Four more than three times a number *x* is twenty-five. Find the number.

74. A number *p* divided by seven minus four is two. Find the number.

75. Twice a number *w* increased by eight is ten. Find the number.

76. Colby rented a beach bike in Seaside. There was an initial fee of $10 and an additional charge of $5 per hour. His total bill was $45. For how many hours did Colby rent the bike? Show all work necessary to justify your answer.

77. The perimeter of a rectangle is 56 feet. The length is 7 feet. What is the width of the rectangle? Explain how you know your answer is correct.

TIC-TAC-TOE ~ EQUATION BOARD GAME

This board game should be created for two people (or two teams) to use. Players will move forward on the path if they solve the given equations correctly. They must solve the final equation in the Winner's Circle to win the game.

To create the game:

Step 1: Write 40 different one- or two-step equations. Find the solution to each equation.

Step 2: Create a board game that has two paths. Each path should have 20 boxes to the Winner's Circle. Place one of your equations in each box.

Step 3: Create a difficult two-step equation for both Winner's Circles.

Step 4: Create an answer key for each path on separate sheets of paper, including the answer to the Winner's Circle equations. The answer keys will be given to the players to check the opponent's answers.

Step 5: Type or write rules for your game. Describe how a player moves towards the Winner's Circle and how they win the game. List other supplies that are necessary, such as a spinner or die.

CAREER FOCUS

MAI
TRAVEL CONSULTANT

I am a Travel Consultant. I arrange and plan travel for people around the world. It is important for me to know about the geography, customs and important features of the many places my customers want to visit. I also have to look for good deals on flights, hotels and other outings people might want to do on vacation. Many travel consultants specialize in one area of travel. Some consultants might know a lot about a certain city or country. Others might focus on certain kinds of trips people take, like cruises. Travel consultants often travel to different parts of the world. Travel consultants can give firsthand information to their customers by going to those places.

Travel consultants use math nearly every day. Some days I use basic math skills to convert money from other countries into American dollars. Other days I may change kilometers to miles or find the percentage of tax on a certain flight. When customers travel to Europe, airports weigh their baggage in kilograms instead of pounds. When I need to convert between two measurement systems, I use math.

Schooling and training is different for each travel consultant. Most consultants need to have a high school diploma and some trade school education. Consultants must also get certified by passing certain tests. It is very helpful for travel consultants to know another language, although it is not required.

Most travel consultants start out making $9.00 to $12.00 dollars per hour. A manager or owner of a travel agency can earn between $35,000 to $45,000 per year. Experienced travel consultants, with lots of world knowledge, can earn up to $70,000 per year.

I love being a travel consultant because it brings me great joy to help a family plan a trip of a lifetime. It is very rewarding to help others have a successful and memorable vacation or trip. I also like how my career allows me to travel around the world and learn a great deal about other people's lifestyles, culture and cuisine.

CORE FOCUS ON INTRODUCTORY ALGEBRA

BLOCK 4 ~ INTEGERS AND FUNCTIONS

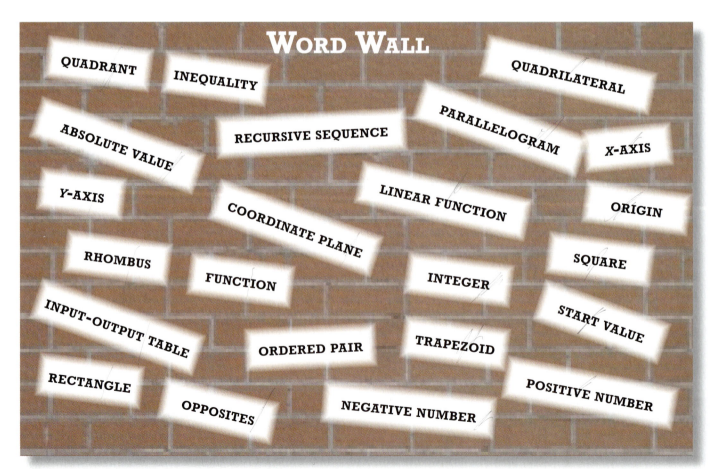

WORD WALL

QUADRANT
INEQUALITY
QUADRILATERAL
ABSOLUTE VALUE
RECURSIVE SEQUENCE
PARALLELOGRAM
X-AXIS
Y-AXIS
LINEAR FUNCTION
ORIGIN
COORDINATE PLANE
RHOMBUS
FUNCTION
INTEGER
SQUARE
INPUT-OUTPUT TABLE
START VALUE
ORDERED PAIR
TRAPEZOID
RECTANGLE
POSITIVE NUMBER
OPPOSITES
NEGATIVE NUMBER

GRAPHING BROCHURE

Create a step-by-step brochure to explain how to make a graph for a given equation.

See page 141 for details.

DOT-TO-DOT

Make dot-to-dot pictures on the coordinate plane. Include directions for connecting the dots using ordered pairs.

See page 120 for details.

PICK A PLAN

Examine three different text messaging plans to find which plan is best for each individual.

See page 141 for details.

CELSIUS TO FAHRENHEIT

Given a conversion table, generate an equation that allows you to convert any temperature in Celsius to Fahrenheit.

See page 136 for details.

POINT OF INTERSECTION

Graph two different functions on the same coordinate plane to determine where the lines cross.

See page 150 for details.

MATH DICTIONARY

Make a dictionary for all the vocabulary terms in this textbook. Create diagrams when possible.

See page 113 for details.

A DAY AT THE BEACH

Write equations and create graphs for different situations that might occur at the beach.

See page 146 for details.

CAREERS USING ALGEBRA

Research and write a report about two different career choices where knowledge of algebra is essential.

See page 130 for details.

MAKING MONEY

Produce graphs that illustrate non-linear situations such as doubling and tripling.

See page 130 for details.

UNDERSTANDING INTEGERS

Understand integers and place integers on a number line.

State	Record Low Temperature	Location	Date
Alaska	−80° F	Prospect Creek Camp	January 23, 1971
California	−45° F	Boca	January 20, 1937
Idaho	−60° F	Island Park Dam	January 18, 1943
Oregon	−54° F	Seneca	February 10, 1933
Washington	−48° F	Winthrop & Mazuma	December 30, 1968

The record low temperatures for five Western states are shown in the table above. The numbers used to represent the low temperatures are called **negative numbers**. Negative numbers are numbers less than 0. **Positive numbers** are any numbers greater than 0.

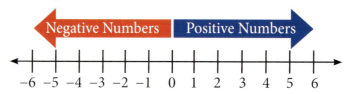

Two numbers are **opposites** if they are the same distance from 0 on a number line, but on opposite sides of 0. The opposite of 4, which is a positive number, is −4. The number −4 is read "negative four". **Integers** are the set of all positive whole numbers, their opposites and zero.

EXAMPLE 1

Find and graph the opposite of each integer.
a. 2　　　　　　　　b. −6

SOLUTIONS

a.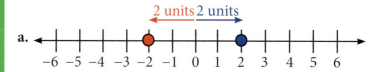

The opposite of 2 is −2 because each integer is 2 units from 0, but in opposite directions.

b.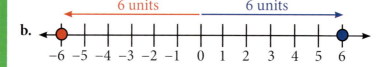

The opposite of −6 is +6 (which is written as 6) because each integer is 6 units from 0, but in opposite directions.

The **absolute value** of a number is the distance that number is from 0 on a number line. The absolute value of −8 is written |−8|. The absolute value of a number is always positive. For example, |−8| = 8.

EXAMPLE 2

Find each absolute value.
a. |−4| **b.** |3|

SOLUTIONS

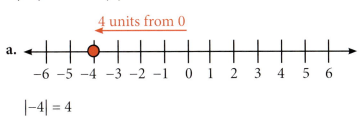

|−4| = 4

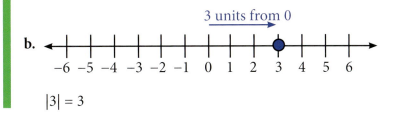

|3| = 3

Integers are used in everyday situations. Positive integers are associated with gains and increases. Negative integers are used when describing losses or decreases. Negative integers are also used to describe something that is below sea level while positive integers describe something above sea level.

EXAMPLE 3

Write an integer to represent each situation.
a. Death Valley, California is 282 feet below sea level.
b. The stock market gained 5 points yesterday.
c. Patrice withdrew $15 from her savings account.

SOLUTIONS

a. Death Valley, California is 282 feet <u>below</u> sea level.
   Answer: −282

b. The stock market <u>gained</u> 5 points yesterday.
   Answer: 5

c. Patrice <u>withdrew</u> $15 from her savings account.
   Answer: −15

EXERCISES

Find the opposite of each number.

1. 8

2. −1

3. −7

4. 1.6

5. −109

6. $\frac{3}{4}$

7. Draw a number line that goes from −5 to 5. Graph the number 2 and its opposite.

8. Draw a number line that goes from −5 to 5. Graph the number −4 and its opposite.

9. Draw a number line that goes from −10 to 10. Graph the following integers on the number line: 2, −5, 9, −1 and 4.

10. Which of the following are integers? 6, −2, $3\frac{1}{3}$, −10, −2.7

11. Name one number that is a negative number but is not a negative integer. Explain your answer.

Find each absolute value.

12. $|9|$

13. $|-1|$

14. $|-14|$

15. $\left|2\frac{1}{4}\right|$

16. $\left|\frac{8}{9}\right|$

17. $|-6.8|$

18. What two numbers both have an absolute value of 7? Explain how you know your answer is correct.

Write an integer to represent each situation.

19. The basement floor is 12 feet below the ground.

20. Three people joined a group for lunch.

21. J.T. owes his friend $8.

22. The average high temperature in Arctic Village, Alaska in February is 6° F below zero.

23. Leticia moved forward 2 spaces on a board game.

24. At 14,411 feet, Mount Rainier is the tallest peak in Washington.

25. Create a situation that would be represented by a positive integer. Explain why a positive integer makes sense for your situation.

26. Create a situation that would be represented by a negative integer. Explain why a negative integer makes sense for your situation.

27. Name the integer represented by each point.

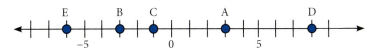

28. On the number line above, which two letters represent numbers that are opposites?

29. A ten-story building has stairs that are each 15 inches tall. Owen, who works on the top floor of the building, says that the change in height for each stair should be represented by −15. Frances, who works on the first floor, thinks that the change in height should be represented by +15. Who do you think is correct? Support your answer with your reasoning.

30. Is the number −4.3 an integer? Why or why not?

REVIEW

Solve each equation using inverse operations. Show all work necessary to justify your answer.

31. $x + 9 = 42$

32. $35 = 7t$

33. $k - \frac{1}{3} = \frac{1}{2}$

34. $14 + a = 91$

35. $\frac{y}{4} = 2.3$

36. $1.4p = 11.2$

37. The area of a rectangle is 32 square inches. The length is 4 inches. What is the width? Show all work necessary to justify your answer.

38. A rectangle has a width of 5.5 centimeters and an area of 21.45 square centimeters. What is the length of the rectangle? Show all work necessary to justify your answer.

39. The base of a triangle is 10 meters. The area of the triangle is 30 square meters. What is the height of the triangle? Explain how you know your answer is correct.

TIC-TAC-TOE ~ MATH DICTIONARY

Create an "Introductory Algebra" Dictionary. Locate all vocabulary words from the four blocks in this textbook. Alphabetize the list of words and design a dictionary. The dictionary should include each word, spelled correctly, along with the definition. If appropriate, diagrams or illustrations should be included.

COMPARING INTEGERS

Compare and order integers.

On a number line, the numbers get larger as you move from left to right. On the number line below, the integer represented by the letter A is the smallest integer marked on the number line. Point E represents the largest integer marked on the number line.

$$A < B < C < D < E$$

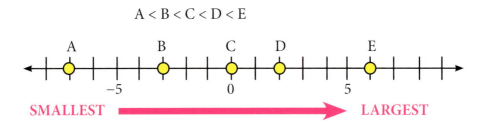

SMALLEST ⟶ LARGEST

EXPLORE!   **WHO'S THE GREATEST?**

Mikhail ran the ball six times in last night's football game. The table shows his yardage on each play.

Play	Yardage
1	9
2	−2
3	5
4	18
5	−7
6	−10

Step 1: During which play did he gain the most yards?

Step 2: Which play was his worst? How do you know?

Step 3: List the yardage amounts from least to greatest.

The scores after the first day of a golf tournament are shown in the table at the right. In golf, the lowest score wins.

Step 4: Who is leading the tournament? Who is in last place?

Step 5: List the players in order from 1st place to 5th place.

Step 6: How many strokes apart are first and last place?

Player	Score
Lamb	−3
Jackson	0
Thompson	−6
Martinez	−4
Kaiser	−1

EXAMPLE 1

Complete each statement using < or >.

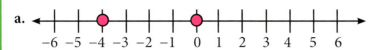

a. −4 ⬤ 0   b. −3 ⬤ −5   c. −6 ⬤ 7

SOLUTIONS

a.

```
←——+——+——●——+——+——+——●——+——+——+——+——+——+——→
  −6 −5 −4 −3 −2 −1  0  1  2  3  4  5  6
```

SMALLEST ➡ LARGEST

Since −4 is to the left of 0 on the number line, −4 < 0.

> Remember:
> < means less than
> > means more than

b. Since −3 is to the right of −5 on the number line, −3 > −5.

c. Because −6 is to the left of 7 on the number line, −6 < 7.

EXAMPLE 2

Seneca, Oregon is located in the mountains of Eastern Oregon. Seneca holds the record for the lowest temperature ever recorded in Oregon at −54° F in February 1933. The table shows the four lowest temperatures on record in Seneca since 1948 (when digital records began). List the temperatures from least to greatest.

Seneca Records since 1948

Date(s)	Lowest Temperature (F°)
January 26, 1957 February 4, 1985	−43
January 22, 1962 December 30, 1978	−41
December 9, 1972 January 1, 1979	−40
December 23, 1983 February 6, 1989	−48

SOLUTION

When the temperatures are graphed on the number line, the lowest temperature will be furthest to the left. The highest temperature will be furthest to the right.

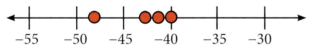

```
←——+——+——●——+——●●●——+——+——→
  −55  −50  −45  −40  −35  −30
```

Listed from least to greatest: −48, −43, −41, −40

EXERCISES

1. Three integers have been graphed on a number line. Explain how you can tell which number is the smallest and which number is the largest by looking at the number line.

Copy and complete each statement using < or >.

2. 9 ⬤ 11

3. 5 ⬤ 0

4. 0 ⬤ −2

5. −8 ⬤ −3

6. −7 ⬤ 4

7. −22 ⬤ −25

Order each set of integers from least to greatest.

8. 3, 0, 4, −1

9. 5, 2, −2, 0, −1

10. −4, −10, 4, −12, −6

11. −3, −9, −12, −1, −5

12. −73, −21, −40, −29

13. 17, 21, −17, −21, 0

14. The table below lists William's last five bank transactions.

a. Copy and complete the table by using an integer to describe each transaction:

Transaction	Integer
Deposited $30	
Withdrew $12	
Withdrew $10	
Deposited $5	
Withdrew $13	

b. List William's transactions from least to greatest.

15. Name five integers between −4 and 4. Order them from least to greatest.

Copy and complete each statement using <, > or =.

16. |6| ⬤ |7|

17. |−4| ⬤ |−5|

18. |8| ⬤ |−8|

19. Jules put a list of numbers in order from least to greatest. His list is below.

$$-1, -3, -7, -10, 0, 1, 4, 9$$

a. Did he order them correctly? If not, explain his mistake.
b. If necessary, correctly order the numbers from least to greatest.

20. Maria scored −7 points in a card game. James had 1 point and Craig had −4 points. The lowest score wins the game. Who won the game? Explain how you know your answer is correct.

REVIEW

For each exercise, find the missing value and state your answer in a complete sentence. Use the formulas in the box at right. See Lesson 3.7 for descriptions of each formula.

21. Chrissy has had 40 'at bats' this season. Her batting average is 0.300. How many hits has she had this season?

22. Leslie went for a walk that lasted 0.75 hours. She covered a total distance of 4.125 miles. At what rate was Leslie walking?

23. Austin put $400 in a savings account that earns simple interest. He earned $40 in two years. What was his interest rate?

Formulas

$$B = \frac{h}{a}$$

$$d = rt$$

$$I = p \cdot r \cdot t$$

THE COORDINATE PLANE

 Graph points on the coordinate plane.

The **coordinate plane** is created by two number lines. The horizontal number line is called the *x*-**axis** and the vertical number line is called the *y*-**axis**. The two number lines cross each other at zero. This point of intersection is called the **origin**.

The *x*-axis and the *y*-axis divide the coordinate plane into four parts. The parts are called **quadrants**. Starting at the top right quadrant and moving counter-clockwise, they are named Quadrant I, Quadrant II, Quadrant III and Quadrant IV.

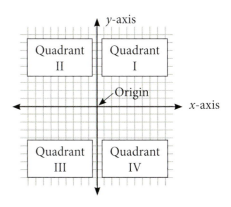

Each point on a coordinate plane is identified by an ordered pair which has two numbers inside parentheses. The first number represents the *x*-coordinate so it relates to the *x*-axis. The second number represents the *y*-coordinate so it relates to the *y*-axis. The origin is the point (0, 0) because it occurs at 0 on both the *x*-axis and the *y*-axis.

EXAMPLE 1

Graph each point on the coordinate plane.
a. A(5, 6)
b. B(0, −3)
c. C(−2, 4)
d. D(−4, −1)

SOLUTIONS

a. Start at the origin. Move right 5 units and up 6 units.

b. Start at the origin. Move 3 units down since the *y*-value is negative. There is no horizontal movement.

c. Start at the origin. Move left 2 units and up 4 units.

d. Start at the origin. Move left 4 units and down 1 unit.

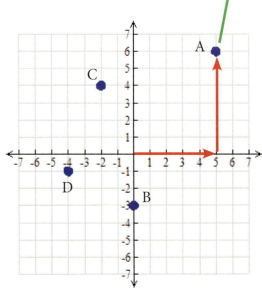

(5, 6)
Start at the origin and move right 5 units and up 6 units.

Two treasure hunters each discovered three different buried treasures in one area of the Pacific Ocean. In order not to forget where the treasures are located, they each created a grid that records the treasures' positions. Each treasure hunter is seeking the other's treasure. If one treasure hunter can locate the other's treasure before his is located, he will get all the treasure. Your goal is to find your opponent's treasure first.

Step 1: Draw two coordinate planes. Label the *x*- and *y*-axes. Number each axis from −3 to 3.

Step 2: Each buried treasure covers three whole number ordered pairs. The treasures can be placed horizontally or vertically. Use the axes if you choose. The treasures must fall in at least three quadrants. Draw three treasures on one of the coordinate planes and record the ordered pairs.

Example:

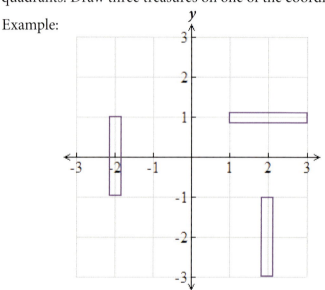

Ordered pairs for example:
(1, 1), (2, 1), (3, 1)
(2, −1), (2, −2), (2, −3)
(−2, 1), (−2, 0), (−2, −1)

Step 3: Pair up with a classmate. Without looking at his or her coordinate plane, guess an ordered pair where you believe a treasure might be. Your partner should reply with "hit" or "miss". Record this guess on your blank coordinate plane. Your partner then takes a turn guessing one point where your treasures might be.

Step 4: A treasure has been "found" when all the ordered pairs have been hit. The person who finds all three treasures first is the winner.

Step 5: Discuss with a classmate how you think ordered pairs might be used in a career that deals with locations in or on the ocean.

EXERCISES

1. Draw a coordinate plane with an *x*-axis and *y*-axis that each go from −5 to 5. Label the *x*-axis, *y*-axis and the origin.

2. Graph and label the following ordered pairs on a coordinate plane.

      A(1, 5)   B(−4, 2)   C(3, 0)   D(0, −2)   E(5, −1)

Give the ordered pair for each point on the coordinate plane shown below.

3. A

4. B

5. C

6. D

7. E

8. F

9. G

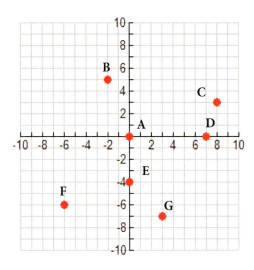

Right or Left then Up or Down

Determine the quadrant each point is in or the axis it lies on.

10. S(4, 9)

11. P(2, 0)

12. L(−3, 2)

13. W(0, −5)

14. H(−1, −7)

15. Z(0.5, 10)

16. M(−5, 0)

17. U(−2, −12)

18. T($5\frac{1}{3}$, $-7\frac{2}{3}$)

19. Quinton graphed the point (2, −3) as shown on the graph at right. Did he graph the point correctly? If not, explain what he did wrong and graph the point in the correct location.

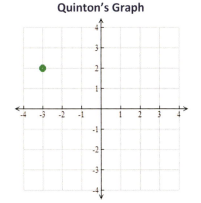

Quinton's Graph

20. When points lie in the same quadrant of a graph, the signs of their x- and y-coordinates can be summarized by the four sign combinations below. Determine which quadrant each sign combination belongs to.
  a. (+, −)
  b. (−, +)
  c. (+, +)
  d. (−, −)

21. Use the points (3, −1), (3, 3), (−2, 3) and (−2, −1).
  a. Graph the points. Connect the points in the order given. Connect the last point to the first point.
  b. Find the perimeter of the figure. Show all work necessary to justify your answer.
  c. Find the area of the figure. Show all work necessary to justify your answer.

22. Use the equation $y = \frac{1}{2}x + 2$. There are pairs of x- and y-values that will make this equation true. Find at least three pairs of values and graph the pairs of values as (x, y) points on a coordinate plane. What do you notice about the graph?

23. Create a shape on a coordinate plane using at least four points. The shape must have two lines of symmetry, the *x*-axis and the *y*-axis.

24. Julie has $1 in an envelope. She wants to double the money in the envelope every week for 5 weeks. Create a graph that shows the amount of money in her envelope for Weeks 0, 1, 2, 3, 4 and 5. Let *x* represent the number of weeks and *y* represent the money in her envelope.

REVIEW

Copy and complete each table by evaluating the given expression for the values listed.

25.

x	$2 + 7x$
0	
1	
2	
3	
5	

26.

x	$\frac{1}{2}x - 1$
2	
6	
7	
10	

Solve each equation. Show all work necessary to justify your answer.

27. $7x = 42$

28. $m - 123 = 968$

29. $\frac{u}{11} = 9$

30. $2.4 = p + 1.7$

31. $z - 5\frac{1}{3} = 12\frac{3}{4}$

32. $7.2 = 1.8a$

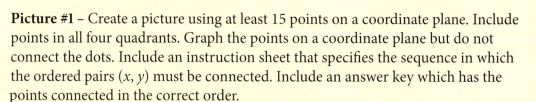

Tic-Tac-Toe ~ Dot-To-Dot

Create two different dot-to-dot pictures.

Picture #1 – Create a picture using at least 15 points on a coordinate plane. Include points in all four quadrants. Graph the points on a coordinate plane but do not connect the dots. Include an instruction sheet that specifies the sequence in which the ordered pairs (*x, y*) must be connected. Include an answer key which has the points connected in the correct order.

Picture #2 – Create a picture using at least 15 points on a coordinate plane. Include points in all four quadrants. DO NOT graph the points on the coordinate plane. Make an instruction sheet that gives the order in which ordered pairs should be connected. Include an answer key that has the ordered pairs graphed and connected to form the picture which you created.

QUADRILATERALS ON THE COORDINATE PLANE

LESSON 4.4

 Use properties of quadrilaterals to solve problems on a coordinate plane.

A quadrilateral is any polygon with four sides. Five common types of quadrilaterals are listed in the table below. Each type of quadrilateral has specific properties that are always true.

Type of Quadrilateral	Definition	Properties
Parallelogram	Any quadrilateral with both pairs of opposite sides parallel.	• Both pairs of opposite sides are parallel. • Both pairs of opposite sides are same length. • Both pairs of opposite angles are congruent. • Consecutive angles add to 180°.
Rectangle	Any quadrilateral with all four angles equal in measure.	• Has all the properties of a parallelogram. • All four angles have same measure (90°).
Rhombus	Any quadrilateral with all four sides equal in measure.	• Has all the properties of a parallelogram. • All four sides are congruent.
Square	Any quadrilateral with four equal sides and four equal angles.	• Has all the properties of a parallelogram. • All four sides are congruent. • All four angles are congruent (90°).
Trapezoid	Any quadrilateral with only one pair of parallel sides.	• Only one pair of opposite sides are parallel. • Angles 1 and 4 add to 180°. (See picture) • Angles 2 and 3 add to 180°

The properties of quadrilaterals can help you determine lengths of missing sides, perimeters and areas. You can use facts about quadrilaterals to find missing points in figures on a coordinate plane. You can also find the lengths of sides on a figure when it is graphed on a coordinate plane.

EXAMPLE 1

Three of the four vertices of a square are located at (1, 2), (4, 2) and (4, 5).
a. What are the coordinates of the missing vertex?
b. Find the area and perimeter of the square.

SOLUTIONS

a. Graph the three points that are given.

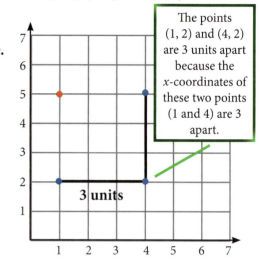

The points (1, 2) and (4, 2) are 3 units apart because the *x*-coordinates of these two points (1 and 4) are 3 apart.

3 units

Every side of a square is equal in length. The length of one of the sides is 3 units, so all sides will be 3 units in length.

The coordinates of the missing vertex must be at (1, 5).

b. Area is found by multiplying length by width. Perimeter is found by adding the lengths of all sides of a figure.

Area = 3 × 3 = 9 units²
3 + 3 + 3 + 3 = 12
Perimeter = 12 units

EXAMPLE 2

Jeannie went on a bike ride around town which was mapped onto a grid where each unit represented one meter in real life. She started at home, which was located at (400, 1500). She biked to the library at (400, 100). Then she went to her friend's house at (1600, 100) where she stayed the night. How far did she bike in all?

SOLUTION

Find the distance between the first two points.

HOME   LIBRARY
(400, 1500) and (400, 100)

The *x*-coordinates are equal. Find the difference between the two *y*-coordinates.

1500 − 100 = 1400

Find the distance between the next two points.

LIBRARY   FRIEND
(400, 100) and (1600, 100)

The *y*-coordinates are equal. Find the difference between the two *x*-coordinates.

1600 − 400 = 1200

Add the distances together.

1200 + 1400 = 2600

Jeannie rode her bike 2,600 meters.

Determine if each statement is ALWAYS TRUE, SOMETIMES TRUE or NEVER TRUE.

1. A square is a rectangle.

2. A rhombus is a square.

3. A rectangle is a parallelogram.

4. A trapezoid is a quadrilateral.

5. Give the name of each quadrilateral and find its area.

a.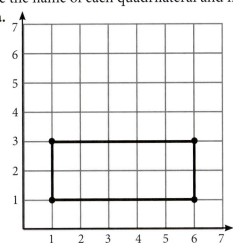

b.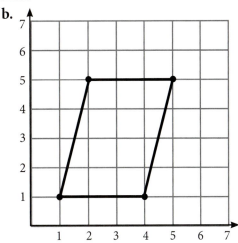

Plot each set of points on a coordinate plane. Connect the points in the order given and connect the last point to the first point. Name the shape with all terms that apply (parallelogram, rectangle, rhombus, square and/or trapezoid).

6. $(0, 4), (0, 0), (4, 0)$ and $(4, 4)$

7. $(-3, 5), (-3, -1), (4, -1)$ and $(4, 5)$

8. $(0, 3), (5, 3), (4, -1)$ and $(-1, -1)$

9. Find the perimeter of each figure in **Exercises 6 and 7**.

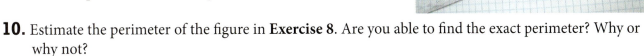

10. Estimate the perimeter of the figure in **Exercise 8**. Are you able to find the exact perimeter? Why or why not?

11. Three of the four vertices of a square are at $(4, 6), (2, 6)$ and $(2, 4)$. What are the coordinates of the missing vertex?

12. Three of the four vertices of a rectangle are at $(1, 5), (1, 3)$ and $(6, 3)$. What are the coordinates of the missing vertex?

13. Three of the four vertices of a parallelogram are at $(0, 0), (-5, 0)$ and $(-7, -3)$. What are the coordinates of the missing vertex? Explain how you know your answer is correct.

14. A lifeguard rotates between four stations on the beach. The beach is laid out on a grid where each unit represents a yard and the four station chairs are located at (30, 100), (30, 180), (100, 180) and (100, 100).

    a. What quadrilateral is formed by the four lifeguard chairs? How do you know?

    b. When the lifeguard does a full rotation of all four chairs and ends up at the chair he started at, how far has he walked? Show all work necessary to justify your answer.

The area of a triangle is found using the formula $A = \frac{1}{2}bh$ where h is the height and b is the base.

15. Graph the triangle formed by the points (5, 1), (1, 1) and (1, 7). What is the area of the triangle?

16. Graph the triangle formed by the points (0, −2), (−5, −2) and (−4, −6). What is the area of the triangle?

The area of a trapezoid is found using the formula $A = \frac{1}{2}h(b_1 + b_2)$ where h is the height of the trapezoid and b_1 and b_2 are the lengths of the top and bottom bases.

17. Graph the trapezoid formed by the points (0, 0), (4, 0), (3, 4) and (1, 4). What is the area of the trapezoid?

18. Graph the trapezoid formed by the points (−1, −4), (7, −4), (5, 3) and (1, 3). What is the area of the trapezoid?

19. Uma went on a nature walk at a park with her family. The map of the park was on a grid where each unit represented a meter. They started at their car, which was parked at (−30, −40).

    a. First they walked to a bird viewing station at (60, −40). How far did they walk?

    b. Next they visited the duck pond at (60, 60). What is the <u>total</u> distance they have walked thus far?

    c. Finally they went to the reptile center located at (−30, 60) and then back to their car. How far did they walk in all? Show all work necessary to justify your answer.

REVIEW

Graph the points below on one coordinate plane. Label each point.

20. M(5, 1)

21. N(−4, −3)

22. P(−1, 2)

23. Q(0, −4)

24. R(2, 0)

25. S(3.5, −3.5)

Evaluate each expression when $a = 5$, $b = 2$ and $c = 10$.

26. $3a + b$

27. $6(c + a)$

28. $c − 2a + 4b$

29. $4bc$

30. $\frac{c}{a} + b$

31. $\frac{4 + c}{b}$

INPUT-OUTPUT TABLES

Create input-output tables for functions.

Andrea reads three books each week during the summer. The equation $y = 3x$ represents the total number of books Andrea has read (y) based on how many weeks have passed (x).

There are times when you are given an equation that describes a relationship between two pieces of information. An equation is useful in creating other ways of displaying the information. The most common ways of displaying data are with graphs, tables and words.

A **function** is a pairing of numbers according to a specific rule or equation. An **input-output table** shows how each input in a function is paired with exactly one output value. The input is usually represented by the variable x and the output is usually the variable y.

Input Weeks, x	Function Rule $y = 3x$	Output Books Read, y
0	3(0)	0
1	3(1)	3
2	3(2)	6
3	3(3)	9

Graphs are another way to visually display equations. The input (x) is paired with the output (y) to create ordered pairs which can be graphed on a coordinate plane.

Input Weeks, x	Function Rule $y = 3x$	Output Books Read, y	Ordered Pair (x, y)
0	3(0)	0	(0, 0)
1	3(1)	3	(1, 3)
2	3(2)	6	(2, 6)
3	3(3)	9	(3, 9)

Label the x- and y- axes so the meaning of the graph is clear.

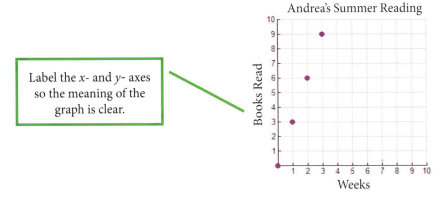

Andrea's book reading has been described using four different mathematical methods:

Words

Andrea reads three
books per week
during the summer.

Mathematical Equation

$y = 3x$ where x represents
weeks and y represents
books read.

Table

Input Weeks, x	Function Rule $y = 3x$	Output Books Read, y
0	3(0)	0
1	3(1)	3
2	3(2)	6
3	3(3)	9

Graph

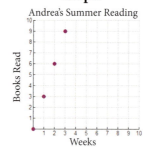

Andrea's Summer Reading

EXAMPLE 1

Complete the input-output table using the function rule given.
Function: $y = 2x + 1$

Input x	Function Rule $y = 2x + 1$	Output y
0		
1		
4		
7		

SOLUTION

Substitute each x-value into the function rule to determine the missing y-values.

Input x	Function Rule $y = 2x + 1$	Output y
0	2(0) + 1	1
1	2(1) + 1	3
4	2(4) + 1	9
7	2(7) + 1	15

Do not forget to use the
order of operations.

Pairing x- and y-values as ordered pairs and graphing these values gives a visual representation of a function. Graphs are used throughout mathematics as a visual model of an algebraic relationship. Look at **Example 2** on the next page to see how a graph can be used as a model for a real-world situation.

EXAMPLE 2

Jennifer received a small kitten for her birthday. The kitten weighed 1 pound. Each week, the kitten gained 0.5 pounds. The weight of the kitten, y, can be found using the function $y = 1 + 0.5x$ where x is the number of weeks she has had the kitten.

a. Complete the table for the first 5 weeks using the function rule.

Input Weeks, x	Function Rule $y = 1 + 0.5x$	Output Weight in pounds, y
1		
2		
3		
4		
5		

b. Write five ordered pairs by pairing the input and output values.
c. Graph the ordered pairs on a coordinate plane.

SOLUTIONS

a. Substitute each x-value into the function rule to determine the missing y-values.

Input Weeks, x	Function Rule $y = 1 + 0.5x$	Output Weight in pounds, y
1	$y = 1 + 0.5(1)$	1.5
2	$y = 1 + 0.5(2)$	2
3	$y = 1 + 0.5(3)$	2.5
4	$y = 1 + 0.5(4)$	3
5	$y = 1 + 0.5(5)$	3.5

b. Write the ordered pairs by pairing each input with the output values:

$(1, 1.5), (2, 2), (3, 2.5), (4, 3), (5, 3.5)$

c. Graph the ordered pairs on a coordinate plane. Since the cat cannot be owned for a negative number of weeks or have a negative weight, you can use the first quadrant only.

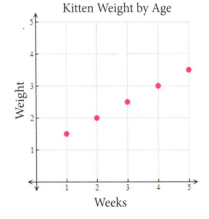

Kitten Weight by Age

EXERCISES

Copy and complete the input-output tables given the function rule.

1.

Input x	Function Rule y = 4x	Output y
0		
1		
2		
3		

2.

Input x	Function Rule y = 3x − 1	Output y
1		
2		
3		
4		

3.

Input x	Function Rule $y = \frac{1}{2}x$	Output y
1		
2		
3		
4		

4.

Input x	Function Rule y = x + 2	Output y
2		
5		
8		
10		

5.

Input x	Function Rule $y = \frac{x}{5}$	Output y
0		
5		
20		

6.

Input x	Function Rule y = 8 − x	Output y
0		
3		
5		
8		

7. Describe in words how to write ordered pairs from a function rule. Write the four ordered pairs from the table in **Exercise 6**.

8. Sonja planted a tomato plant in May. She measured the plant's height each week. She found the height of the plant could be represented by the equation $y = 0.5x + 4$ where x represents the number of weeks that have passed and y represents the height of the plant in inches.

a. Copy and complete the table to show the height of the plant throughout the summer.

Input Weeks, x	Function Rule y = 0.5x + 4	Output Height, y
0		
4		
7		
10		

b. Write the four ordered pairs from the table above by pairing the input and output values.

c. Draw a coordinate plane with an x-axis and y-axis that go from 0 to 10. Label the x-axis as "Weeks" and the y-axis as "Height". Graph the ordered pairs on a coordinate plane.

d. Connect the points with a line. This line shows the continuous growth of the tomato plant.

9. Serj is able to bike 15 miles per hour on the road. He wants to figure out how many miles (y) he can bike based on the number of hours (x) he bikes. He developed a function rule to help him:

a. Copy and complete the table.

Input Hours, x	Function Rule $y = 15x$	Output Miles, y
1		
2		
5		
6		

b. Write the four ordered pairs from the table above by pairing the input and output values.

c. Draw a coordinate plane with an x-axis that goes from 0 to 10 and a y-axis that goes from 0 to 100 as seen below. Graph the ordered pairs on the coordinate plane.

d. Connect the points with a line. This line shows Serj's continuous progress on his bike ride.

e. Serj went for a 12-hour bike ride. How far should he expect to travel on his bike in those twelve hours? Use words and/or numbers to show how you determined your answer.

10. Samantha saves $5 each week. Let x represent the number of weeks she has been saving money and y represent the total amount of money she has saved.

a. Write a function rule for this situation.

b. Create an input-output table for x = 0, 1, 2, 3 and 4. Complete the table.

c. Write the five ordered pairs from the table.

d. Draw a coordinate plane and graph the five points.

REVIEW

Solve each equation for x. Show all work necessary to justify your answer.

11. $x + 7 = 23$

12. $\frac{x}{5} = 3$

13. $x - 2 = 8$

14. $4x = 12$

15. $x + 6.9 = 10$

16. $\frac{x}{9} = 0.1$

17. $x - \frac{1}{3} = \frac{1}{4}$

18. $938 = x + 151$

19. $22x = 77$

TIC-TAC-TOE ~ MAKING MONEY

Non-linear functions are functions that do not form a line when graphed. There are many non-linear functions. An exponential function is a type of non-linear function. Below are two situations that involve exponential functions.

Dawn's neighbor offers her a job. She will be paid $5 for the first week of babysitting. Each week her pay will double.

1. Copy the chart and fill in the missing values.

2. Draw a coordinate plane and graph the ordered pairs represented by the values in the table.

3. Predict how many weeks it will take before she is making more than $1,000.

Week x	Pay y
1	$5
2	
3	
4	
5	
6	

Tom's neighbor offers to pay him $2 to mow his yard the first week of the summer. Each week after that, his neighbor promises to triple his pay.

4. Copy the chart and fill in the missing values.

5. Draw a coordinate plane and graph the ordered pairs represented by the values in the table.

6. Predict how many weeks it will take before he is making more than $1,000 to mow the yard.

Week x	Pay y
1	$2
2	
3	
4	
5	
6	

TIC-TAC-TOE ~ CAREERS USING ALGEBRA

There are many career choices where knowledge of algebra is essential. Research at least two different careers that require the use of algebra. Write a report about these careers.

Include the following components for each career:
  ◆ a description of the career
  ◆ how the career includes the use of algebra
  ◆ how much schooling is required for the career

WRITING FUNCTION RULES

Write function rules for tables or graphs.

In **Lesson 4.5** you were given function rules as mathematical equations. You were asked to make tables and graphs for the function rules. In this lesson, you will learn how to take a table or a graph and develop the function rule for the table or graph.

EXPLORE!   FUNCTION RULES

Justin completes homework assignments for math every week. The graph shows the total number of homework assignments (y) Justin has completed based on the number of weeks (x) he has been in school this year.

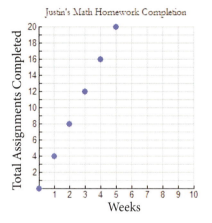

Step 1: Copy and complete the input-output table shown below with the six ordered pairs shown on the graph.

Step 2: When the input column goes up by one each time, it can be called the counting column. Is the input column in this table a counting column?

Step 3: The function rule will have a start value. The start value is found by locating the output value that is paired with the input value of zero. What is the start value for Justin's homework chart? Fill this start value into the developing function rule.

$$y = \_\_\_\_\_$$

Input Weeks, x	Output Assignments Completed, y
0	
1	
2	
3	
4	
5	

Step 4: Look at the output column. Is this column increasing or decreasing as the counting column gets bigger? How much is it increasing or decreasing during each step?

Step 5: If a table has a counting column, the amount of increase or decrease between consecutive outputs is called the amount of change. The amount of change is always the coefficient of x in the function rule. Fill in the amount from **Step 4** into the function rule. (NOTE: Choose the addition symbol if the output column increases and the subtraction symbol if the output column decreases.)

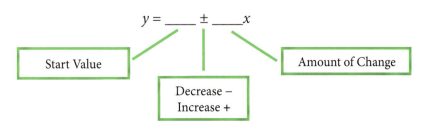

$$y = \_\_\_\_ \pm \_\_\_\_ x$$

Start Value

Decrease –
Increase +

Amount of Change

Step 6: In some cases you do not need to write the start value in the equation because it does not affect the equation. Is this one of those situations? How do you know?

Step 7: Use your function rule to help predict how many homework assignments Justin will complete during the entire school year. There are 36 weeks in the school year.

WRITING FUNCTION RULES FROM TABLES AND GRAPHS

1. If given a graph, create an input-output table using the ordered pairs of the points on the graph.
2. Make sure the input column starts at zero and counts up by one.
3. Determine the start value (the output number paired with an input value of zero).
4. Determine the amount of change (the amount the output numbers are increasing or decreasing by).
5. Write the function rule by filling in the start value, operation and amount of change.

$$y = \underline{\hspace{1cm}} \pm \underline{\hspace{1cm}} x$$

EXAMPLE 1

Find the function rule for the input-output table.

Input x	Output y
0	2
1	5
2	8
3	11

SOLUTION

The input column does start at zero and counts up by one.

The start value is the output value paired with zero.
The start value is 2.

Start the function rule. $\qquad y = 2 \pm \underline{\hspace{1cm}} x$

The amount of change is determined by the amount the output column increases or decreases each step. In this function, the amount of change is increasing by 3.

Input x	Output y	
0	2	
1	5	+ 3
2	8	+ 3
3	11	+ 3

Start Value

Amount of Change

Finish the function rule. $\qquad y = 2 + 3x$

EXAMPLE 2

Find the function rule for the graph.

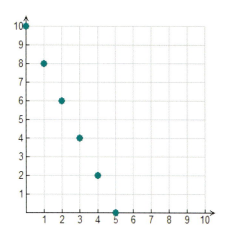

SOLUTION

Create an input-output table using the *x*- and *y*-values from each point.

Input	Output		
x	*y*		
0	10		Start Value
1	8	> – 2	
2	6	> – 2	
3	4	> – 2	
4	2	> – 2	
5	0	> – 2	

Start with the ordered pair that has an *x*-value of zero.

Amount of Change

The input column does start at zero and counts up by one.
The start value is the output value paired with zero.
The start value is 10.

Start the function rule.

$$y = 10 \pm \underline{\quad} x$$

The amount of change is determined by the amount the output column increases or decreases each step.

The amount of change is decreasing by 2.

Complete the function rule.

$$y = 10 - 2x$$

Find the function rule for each input-output table.

1.

Input x	Output y
0	5
1	9
2	13
3	17

2.

Input x	Output y
0	11
1	9
2	7
3	5

3.

Input x	Output y
0	2
1	2.5
2	3
3	3.5

4. Use the function rule $y = 3 + 5x$ to answer the questions.

    a. What is the amount of change? How do you know?

    b. What is the start value?

    c. Is the graph of this function increasing or decreasing? How do you know?

Find the function rule for each graph.

5.

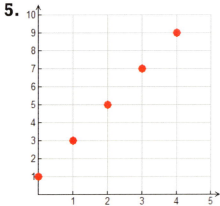

6.

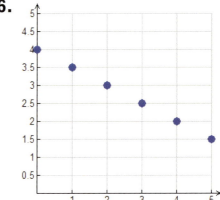

7.

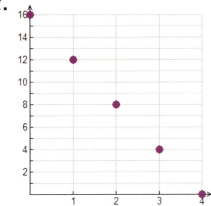

8.

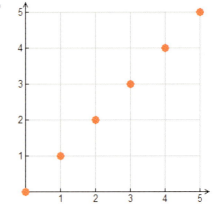

9. Wally spent the summer fighting fires. He started by fighting 12 fires during the training. Each week during the summer he fought an additional 3 fires. He kept track of the total number of fires he helped fight.

    a. Copy and complete the input-output table based on the information above.

Input Weeks, x	Output Total Fires Fought, y
0	12
1	
2	
3	
4	
5	

    b. Find the function rule for the table.

    c. What is the total number of fires Wally fought during his training and the eleven-week fire season? Show all work necessary to justify your answer.

10. Explain the first step to finding a function rule when given a graph of the function.

11. Nanette came up with function rules for two different problems. Her teacher told her the rules were correct but could both be written in a simpler form. Write each answer in simplest form.

    Function Rule #1: $y = 0 + 4x$

    Function Rule #2: $y = 3 + 1x$

12. Dajan started the school year with $350 which he had saved from his summer job. Each week he spent $20 of his savings for snacks and movies.

Input Weeks, x	Output Money Remaining, y
0	
1	
2	
3	
4	
5	

    a. Copy and complete the input-output table based on the information above.

    b. Find the function rule for the table.

    c. How much money would Dajan have left after 8 weeks of school? Explain how you know your answer is correct.

13. Edna runs a drop-off baby-sitting service. She provides a graph to parents that shows their charges for a given number of hours.

    a. What is the initial cost (or start value) to drop a child off?

    b. What is the additional fee per hour for baby-sitting?

    c. Write a function rule that expresses the total cost (y) in terms of the number of hours a child is there (x).

    d. One parent left his child for 9 hours. How much was his total charge for baby-sitting? Show all work necessary to support your answer.

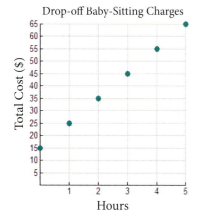

Drop-off Baby-Sitting Charges

14. In this lesson, all the functions you examined and created are linear functions, which means their graphs form lines. Graph the function $y = 2^x$ for the x-values 0, 1, 2, 3, 4 and 5. Is this graph a linear function? Explain your reasoning.

15. Write a function rule that has an ordered pair of (1, 7) and another ordered pair of (4, 16). Use words and/or numbers to show how you found the function rule.

REVIEW

Use the order of operations to evaluate.

16. $5 + 7 \cdot 2 - 3$

17. $\dfrac{4 + 11}{4 - 1}$

18. $(3 + 4)^2 - 6$

19. $10 - 2 \cdot 3 + 7 \cdot 4$

20. $3(2 + 3)^2$

21. $\dfrac{5 + 2 \cdot 14}{3} - 10$

Complete each statement using < or >.

22. 4 ⬤ −9

23. −2 ⬤ 0

24. 5 ⬤ 1

25. −3 ⬤ 3

26. −5 ⬤ −6

27. −101 ⬤ −89

TIC-TAC-TOE ~ CELSIUS TO FAHRENHEIT

Lara went to Canada on vacation. All of the local news reports gave the weather in degrees Celsius. She wanted to convert the temperatures from Celsius to Fahrenheit. She located a conversion table for 0° through 6° Celsius.

1. Find the start value and amount of change.

2. Write a conversion equation using Celsius for x and Fahrenheit for y.

3. Use your conversion equation to determine the Fahrenheit equivalent for 10° C, 20° C and 30° C.

4. Create a graph that shows the linear relationship.

Celsius (C)	Fahrenheit (F)
0°	32°
1°	33.8°
2°	35.6°
3°	37.4°
4°	39.2°
5°	41°
6°	42.8°

GRAPHING LINEAR FUNCTIONS

 Graph linear functions using input-output tables or amounts of change.

A linear function is a function whose graph is a straight line. The function rules you developed in **Lesson 4.6** were for linear functions. Linear functions can be graphed using the information from their function rules. The start value tells you where to start the line on the *y*-axis. The amount of change tells you whether or not you should increase or decrease from the start value and how big each step should be.

Once the points of the function are graphed, a straight line should be drawn through all the points to show the function is continuous. A linear function is continuous because there are an infinite number of points that make the equation true.

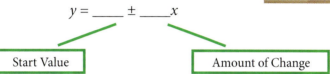

$$y = \underline{\quad} \pm \underline{\quad} x$$

| Start Value | Amount of Change |

EXPLORE!   **FUNCTION FUN**

Step 1: Four linear functions are listed below. Determine the start value and amount of change for each function.

Function 1: $y = 5 + 0.5x$ Function 2: $y = 9 - 2x$
Function 3: $y = 3x$ Function 4: $y = 4 - x$

Step 2: Which functions are increasing? How do you know?

Step 3: There are two methods for graphing linear functions. The first method is to create an input-output table, graph the ordered pairs and connect the points with a line. Try this method with Function 1.

Copy and complete the input-output table.

| Input | Function Rule | Output | Ordered Pair |
x	$y = 5 + 0.5x$	*y*	*(x, y)*
0			
1			
2			
3			

Graph the ordered pairs on a coordinate plane. Draw a line through the points. Put arrows on both ends of the line to show that it is continuous.

Step 4: The second method for graphing a linear function requires using the start value and amount of change to create the graph. Try this method with Function 2: $y = 9 - 2x$. Draw a coordinate plane. Put a point at the location of the start value on the *y*-axis.

Use the amount of change to locate the next point. Always move to the right one unit. If the amount of change is increasing, go up that amount. If the amount of change is decreasing, go down that amount. Put a point at each step.

Continue that process for at least three steps so that you have four points. Draw a line through the points. Put arrows on both ends of the line to show that it is continuous.

Step 5: Choose the method you like best to graph Function 3 and Function 4.

Step 6: Explain to a classmate why you chose the method you did. Ask them which method they chose and why.

EXAMPLE 1

Graph the linear function $y = 10 - 3x$ using input-output tables.

SOLUTION

Use 0 through 3 as your input values when creating a table to help you graph a linear function.

Input x	Function Rule $y = 10 - 3x$	Output y	Ordered Pair (x, y)
0	$10 - 3(0)$	10	$(0, 10)$
1	$10 - 3(1)$	7	$(1, 7)$
2	$10 - 3(2)$	4	$(2, 4)$
3	$10 - 3(3)$	1	$(3, 1)$

Graph the points on a coordinate plane.

Draw a line through the points. Put arrows on the ends of the line.

EXAMPLE 2

Graph the following linear functions using start values and amounts of change.
a. $y = 1 + 2x$ **b.** $y = 8 - x$

SOLUTIONS

a. Start Value = 1
Amount of Change is increasing by 2

Put a point at the location of the start value on the *y*-axis of a coordinate plane.

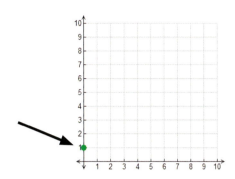

Use the amount of change to locate the next point. Move up 2 units and one unit to the right since the amount of change is increasing by 2. Continue this so you have a total of 4 points. Draw a line through the points. Put arrows on both ends of the line to show that it is continuous.

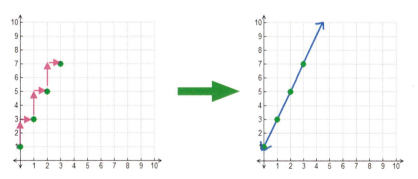

b. Start Value = 8
Amount of Change is decreasing by 1

> If a variable does not have a front coefficient, the coefficient is 1.

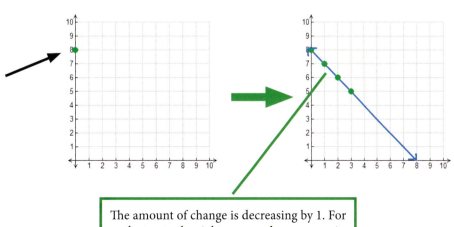

> The amount of change is decreasing by 1. For each step to the right, you go down one unit.

EXERCISES

Graph each linear function using input-output tables.

1. $y = 1 + 3x$

2. $y = 8 - 2x$

3. $y = 1.5x$

4. $y = 3 + x$

5. $y = 7 - \frac{1}{2}x$

6. $y = 15 - 4x$

Graph each linear function using start values and amounts of change.

7. $y = 5 + 2x$

8. $y = 9 - 4x$

9. $y = 5 - x$

10. $y = 3x$

11. $y = 4 + \frac{1}{2}x$

12. $y = 8 - 3x$

13. Graph $y = 10 - 2x$ using either method (input-output tables or start value/amount of change). Explain why you chose the method you did.

14. Perry's little sister weighed 8 pounds when she was born. She gained approximately 0.5 pounds each week. Let x represent the age of the baby in weeks and let y represent the baby's weight in pounds.
   a. What is the start value?
   b. What is the amount of change? Is it increasing or decreasing?
   c. Write a function rule that models this situation.
   d. Graph the linear function on a coordinate plane.
   e. Do you think the baby will continue to grow at this rate until she is an adult? Why or why not?

15. Tiara planted a sunflower in her garden. The flower was 9 inches tall when she planted it. It grew 3 inches each week. Let x represent the number of weeks since she planted the flower and let y represent the flower's height. Write a function rule for the flower's growth and graph the function.

16. Skyler's total savings are shown on the graph.
   a. Find the function rule that represents the graph.
   b. How much will Skyler have in his savings account after 18 months? Explain how you determined your answer.

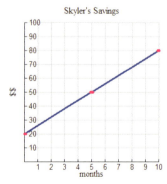

17. Three linear equations have a unique relationship. Their equations are $y = 2 + 3x$, $y = 5 + 3x$ and $y = 3x$. What is their relationship? Use words and/or numbers to show how you determined your answer.

Write the function rule for each graph.

18.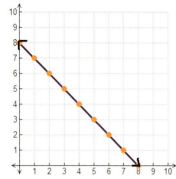

19.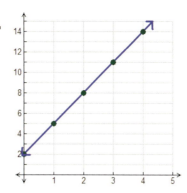

20.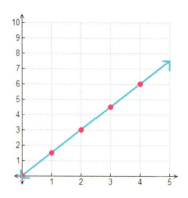

TIC-TAC-TOE ~ PICK A PLAN

Digital Groove is a website that allows members to download songs to their devices. They have three different downloading plans. Three friends are trying to choose the right plan based on how much they will use the service. Nigel estimates that he would download about 200 songs per year. Heidi thinks she will only download about 70 songs per year. Brandon downloads well over 1000 songs per year.

DOWNLOAD PLAN OPTIONS:
Option 1: $15 yearly fee plus $0.02 per song download
Option 2: $8 yearly fee plus $0.10 per song download
Option 3: $0.15 per song download

Write a linear function for each download option. Let *x* equal the number of song downloads and *y* represent the total cost per year. Determine which plan is the best option for each of the three friends. Support your answers with mathematics.

TIC-TAC-TOE ~ GRAPHING BROCHURE

Two different methods are shown for graphing a linear function in **this lesson**. Create a brochure that describes each process step-by-step. Use graphs, tables and illustrations. Include an example for each method.

PATTERNS AND FUNCTIONS

Write function rules for patterns.

A recursive sequence uses a first term and a repeated operation to create an ordered list of numbers.

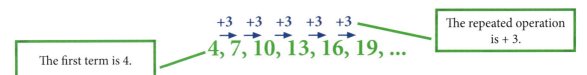

+3 +3 +3 +3 +3

4, 7, 10, 13, 16, 19, ...

The first term is 4.

The repeated operation is + 3.

EXAMPLE 1 State the value of the first term, the operation and the next three terms for each of the following recursive sequences.

a. 8, 17, 26, 35, 44, …   b. 26, 24, 22, 20, 18, …

SOLUTIONS
a. First term = 8
Operation = + 9
Next three terms: 53, 62, 71

b. First term = 26
Operation = − 2
Next three terms: 16, 14, 12

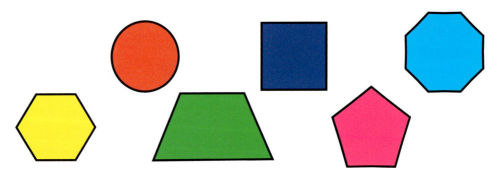

Many patterns in geometry are recursive sequences. It is important to examine the recursive sequence by writing out the list of numbers in the sequence when working with patterns.

EXAMPLE 2 Squares, each 1 unit by 1 unit, are placed next to each other one at a time to form a long strip of squares.

Stage 0   Stage 1   Stage 2

a. What is the perimeter of Stage 0 using just one square?
b. What is the perimeter of the figure using two squares? Three squares?
c. Draw the next two stages in the pattern. What are the perimeters of these figures?
d. What is the operation that describes the change in the perimeter from one stage to the next?
e. Predict the perimeter of the figure in Stage 9.

EXAMPLE 2
SOLUTIONS

a. Each side of the square is 1 unit. There are four sides on the first square. The perimeter of the first square is 4 units.

b. The perimeter of Stage 1 using two squares is 6 units. The Stage 2 figure has a perimeter of 8 units.

c. The next two figures in the pattern are:

Perimeter = 10 units   Perimeter = 12 units

d. The operation is the amount added from one stage to the next.

Operation = +2

e. The perimeter of Stage 9 in this sequence can be found by listing out the terms.

4, 6, 8, 10, 12, 14, 16, 18, 20, 22

Stage 9

Perimeter = 22 units

Create a visual representation of the pattern in **Example 2** by graphing the values on a coordinate plane. The numbers along the x-axis represent the stage number. The numbers along the y-axis represent the perimeter.

In order to find the function rule, you need to know the start value and the amount of change. The start value for this linear function is 4 because that is where the graph begins on the y-axis. The repeated operation or amount of change is +2. Determine the function rule using the method from earlier in this Block.

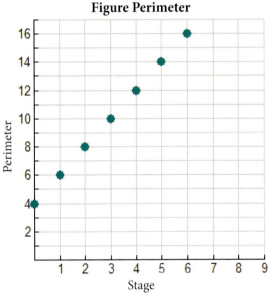

Figure Perimeter

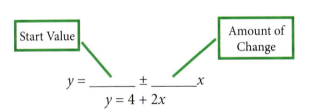

Start Value   Amount of Change

$$y = \underline{\quad} \pm \underline{\quad} x$$
$$y = 4 + 2x$$

This linear function can be used to determine the perimeter of any term in this pattern. For example, to determine the perimeter of Stage 20, substitute 20 into the function for x.

$$y = 4 + 2(20)$$
$$= 4 + 40$$
$$= 44$$

Notice that a line was not drawn through the points on the graph above. This linear function is not continuous because you only have whole number stages for the x-values. For example, there is not a Stage 1.5 or 6.4. When graphing a linear function, carefully consider whether or not a line should be drawn through the points on the graph.

Copy each recursive sequence. Fill in the missing values. Identify the first term and the operation that must be performed to arrive at the next term.

1. 3, 5, 7, 9, ____, ____, ____

2. 27, 22, 17, ____, ____, ____

3. 125, 150, ____, 200, ____, ____

4. 23, ____, 29, 32, ____, ____

5. ____, 8.6, 8.2, 7.8, ____, ____

6. 4, $5\frac{1}{2}$, ____, ____, 10, ____

7. Triangles, each side one unit in length, are placed together (one at a time) to form a long strip of triangles.

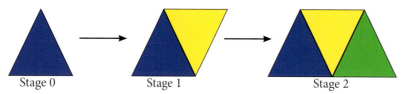

Stage 0   Stage 1   Stage 2

    a. What is the perimeter of Stage 0 using just one triangle?
    b. What is the perimeter of the figure using two triangles? Three triangles?
    c. Draw the next two stages in the pattern. What are the perimeters of these figures?
    d. What operation describes the perimeters from one stage to the next?
    e. Predict the perimeter of the figure in Stage 8. Explain how you know your answer is correct.

8. Use the pattern below to answer the questions.

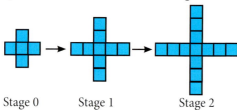

   Stage 0   Stage 1   Stage 2

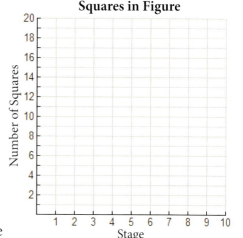

    a. How many squares are used in Stage 0? In Stage 1? In Stage 2?
    b. Draw the figure that represents the next stage. List how many squares are used in this stage.
    c. What operation describes the number of squares from one stage to the next?
    d. Draw a coordinate plane similar to the one at right. Graph the first five stages.
    e. Write a function rule for this pattern that will determine the number of squares in a figure when given the stage number.
    f. How many squares would be used in Stage 15? Use words and/or numbers to show how you determined your answer.

9.

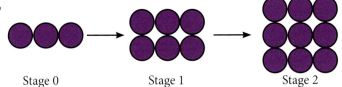

Stage 0   Stage 1   Stage 2

a. How many circles are used in Stage 0? In Stage 1? In Stage 2?
b. Draw the figures that represent the next two stages. List how many circles are used in each.
c. What is the operation that describes the number of circles from one stage to the next?
d. Draw a coordinate plane and graph the first five stages.
e. Write a function rule that will determine the amount of circles in a figure when given the stage number.
f. How many circles would be used in Stage 30? Use words and/or numbers to show how you determined your answer.

10. Generate your own recursive sequence.
a. Describe the recursive sequence by stating the first term and the repeated operation.
b. List out the first five numbers in the sequence.
c. What is the 15th term in your sequence? Show all work necessary to justify your answer.

11. The table below shows the number of triangles in each stage of a geometric pattern. How many triangles would be used in the 100th stage? Use words and/or numbers to show how you determined your answer.

Stage 0	Stage 1	Stage 2	Stage 3	Stage 4
6	13	20	27	34

REVIEW

Graph the following linear functions using the input-output tables method OR the start value and amount of change method.

12. $y = 3 + 2x$ **13.** $y = 14 - 3x$ **14.** $y = 0.5x$

15. $y = 4 + x$ **16.** $y = 9 - \frac{1}{2}x$ **17.** $y = 9 - 2x$

18. Willy determined that the amount in his savings account can be modeled by the function rule $y = 50 + 10x$. The variable y is the amount in his account. The variable x is the number of weeks that have passed.
a. How much did he start with in his account?
b. How much does he save every week?
c. Graph the linear function. Include points through the first five weeks.

19. Write the function rules for the graphs below.

a.

b.

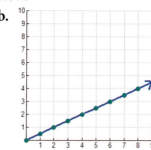

20. Copy and complete the input-output table for the function rule.

Input, x	Function Rule $y = 124 - 1.7x$	Output, y
2.5		
5		
8.3		
10.8		

Tᴵᴄ-Tᴬᴄ-Tᴼᴱ ~ A Dᴬʸ ᴬᵀ ᴛʜᴱ Bᴱᴬᴄʜ

The Larson family went to the Gulf Coast and noticed algebraic situations all around them. Examine two situations they encountered and answer the given questions.

Situation #1 – Body Board Rentals
The local surf shop charges an initial fee of $5 for a body board. There is an additional fee of $3 per hour.

1. Write a function rule for this situation where x represents the number of hours and y represents the total cost for the rental.
2. Assume the situation creates a continuous graph. Graph the function rule that shows the cost for 0 through 5 hours.
3. What is the cost of a 4 hour body board rental?

Situation #2 – Ski Ball Tickets
The Larson kids found out they can win tickets at the ski ball arcade. The ski ball machine automatically gives each player four tickets. The machine will give the player one additional ticket for each ball that goes in the center target.

4. Create a table that shows the number of tickets received based on the number of balls that land in the center target. Include getting 0 through 6 balls in the center target.
5. Write a function rule for this situation where x represents the number of balls that go into the center target and y represents the total number of tickets received.
6. How many tickets did one child receive if he got 9 balls in the center target?

INEQUALITIES

Write inequalities for real-world situations.
Graph inequalities on a number line.

Jennifer has less than $75 in her bank account. Monte has at least 300 baseball cards in his collection. Each of these statements can be written using an **inequality**. Inequalities are mathematical statements which use >, <, ≥ or ≤ to show a relationship between quantities.

Jennifer has less than $75 in her bank account.   Monte has at least 300 baseball cards.

$$j < \$75 \qquad\qquad m \geq 300$$

Inequalities have multiple answers that can make the statement true. In Jennifer's example, she might have $2 or $74. All that is known for certain is that she has under $75 in her account. In Monte's example, he might have 300 cards (this comes from the "equal to" part of ≥) or 3,000,000 cards. There is an infinite number of possibilities that would make this statement true.

INEQUALITY SYMBOLS

> "greater than"
< "less than"
≥ "greater than or equal to"
≤ "less than or equal to"

EXAMPLE 1

Write an inequality for each statement.
a. Nina's height (h) is greater than 50 inches.
b. Pete has at least $6 in his pocket. Let p represent the amount of money.
c. Jeremy is less than 20 years old. Let j represent Jeremy's age.

SOLUTIONS

a. The key words are "greater than". Use the > symbol. $h > 50$

b. The key words are "at least". This means the lowest
   amount he could have is $6.00. Use the ≥ symbol. $p \geq 6$

c. The key words are "less than". Use the < symbol. $j < 20$

The solutions to an inequality can be represented on a number line. Each inequality has an infinite number of solutions. The number line diagrams below show a visual representation of each inequality statement. The open or closed circle shows whether or not the starting point is included in the solution. A closed (or filled in) circle includes that value while an open circle does not include the value. An arrow is used to indicate that the solutions continue indefinitely in that direction.

$x > -1$

x could be any value greater than −1.

$x < -1$

Notice that the open circles on the number line DO NOT include that value while filled in circles DO include the value.

$x \geq -1$

$x \leq -1$

EXAMPLE 2

Graph each inequality on a number line.
a. $x < 5$ b. $y \geq 0$

SOLUTIONS

a. This inequality includes all values less than 5 (not including 5). Use an open circle on 5 to show this. The arrow points to all values less than 5.

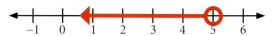

b. This inequality includes all values greater than or equal to 0. Use a closed circle on 0 to show this. The arrow points to all values greater than 0.

EXERCISES

Write an inequality for each graph shown. Use x as the variable.

1.

2.

3.

4.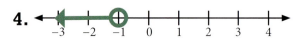

Write an inequality for each statement. Graph each inequality on a number line.

5. Ivan's shoe size (*s*) is greater than or equal to 7.

6. An ice cream cone costs more than $2.50. Let *c* represent the cost.

7. A travel company is offering cruises for less than $500. Let *t* represent the cost of the cruise.

8. Hamburgers at Mel's Diner have at least 400 calories. Let *h* represent the calories in a hamburger.

9. Jennifer ran no more than 50 miles this week. Let *m* represent the number of miles she ran.

Match each statement on the left to its corresponding number line. Number lines may be used more than once.

10. $x \geq 2$

11. Matt has less than 2 pounds of candy.

12. A chicken weighs at least 2 pounds.

13. $x \leq 2$

14. Wendy spent more than $2 on lunch.

15. Trent ran at most 2 miles.

16. $x < 2$

A.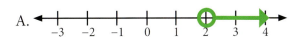

B.

C.

D.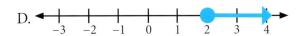

17. Michael says he doesn't understand the difference between the symbols > and ≥. Explain the difference between the two symbols in words.

18. Some number line diagrams are defined by two inequalities. Write two inequalities (using "and" between the two statements) to describe each number line shown.

a.

b.

19. Wilson told Joey, "There are an infinite number of values for *x* that make the statement *x* < 10 true." Joey disagreed. Joey said the only values that make that statement true are 1, 2, 3, 4, 5, 6, 7, 8 and 9. Who do you agree with? Why?

Solve each equation for the variable. Show all work necessary to justify your answer.

20. $2x + 5 = 17$

21. $\frac{y}{3} - 1 = 3$

22. $5 + \frac{m}{10} = 9$

23. $10x - 9 = 81$

24. $8 = \frac{y}{5} + 7$

25. $273 + 51x = 375$

Write an equation for each statement. Solve each problem and check your solution.

26. Three times a number x minus two is twenty-five. Find the number.

27. The sum of ten and the number m divided by six is thirteen. Find the number.

28. Twenty increased by seven times a number w is ninety-seven. Find the number.

29. Judy ran six less than four times the number of miles that Pete ran. Judy ran forty-two miles. How many miles did Pete run? Use words and/or numbers to show how you determined your answer.

30. The perimeter of a rectangle is 125 inches. The width is 37 inches. What is the length? Show all work necessary to justify your answer.

TIC-TAC-TOE ~ POINT OF INTERSECTION

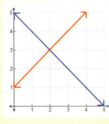

A set of two or more linear equations is called a system of linear equations. The solution to the system of linear equations is found at the point of intersection of the two lines. It is the ordered pair (x, y) that makes both linear equations true.

Given two linear equations, you can find the point of intersection by carefully graphing the two functions and determining the point where the two lines cross.

For each problem below:
a. Graph the pair of linear functions on the same coordinate plane.
b. Find the solution to the system of equations by locating the point of intersection. Give the exact point of intersection as an ordered pair.
c. Show that the solution works in both equations by substituting it for x and y to see if each equation is true.

1. $y = 1 + 2x$
$y = 10 - x$

2. $y = 8 - 2x$
$y = 3 + 0.5x^{-}$

3. $y = 10 - 3x$
$y = 2x$

Vocabulary

absolute value	opposites	rectangle
coordinate plane	ordered pair	recursive sequence
function	origin	rhombus
inequality	parallelogram	square
input-output table	positive number	start value
integer	quadrant	trapezoid
linear function	quadrilateral	*x*-axis
negative number		*y*-axis

 Understand integers and place integers on a number line.
Compare and order integers.
Graph points on the coordinate plane.
Use properties of quadrilaterals to solve problems on a coordinate plane.
Create input-output tables for functions.
Write function rules for tables or graphs.
Graph linear functions using input-output tables or amounts of change.
Write function rules for patterns.
Write inequalities for real-world situations.
Graph inequalities on a number line.

Lesson 4.1 ~ Understanding Integers

Find the opposite of each number.

1. 6

2. −9

3. −11

4. Draw a number line that goes from −5 to 5. Graph the number −3 and its opposite.

Find each absolute value.

5. |7|

6. |−3|

7. |−46|

Write an integer to represent each situation.

8. The buried treasure is 18 feet below the ground.

9. Cameron owes his mother $5.

10. Quinn's stock increased by $9.

Copy and complete each statement using < or >.

11. 9 ⬤ 11

12. 5 ⬤ 0

13. 0 ⬤ −2

Order the integers from least to greatest.

14. 5, 0, −4, 1

15. 7, −3, −2, 0, 2

16. −1, −9, −6, −12, −3

17. Use the table below.
    a. Copy and complete the table by using an integer to describe each week's activity.
    b. List the integers that represent the stock activity from least to greatest.

Technogram Stock Account Activity

Week	Activity	Integer
1	Increased by 5 points	
2	Dropped 3 points	
3	Gained 10 points	
4	Fell 7 points	
5	Decreased by 6 points	

Lesson 4.3 ~ The Coordinate Plane

18. Draw a coordinate plane with an *x*-axis and *y*-axis that each go from −10 to 10. Label the *x*-axis, *y*-axis and the origin. On the coordinate plane, graph and label the following ordered pairs:
    R(4, 3)   S(0, −2)   T(−9, 7)   U(6, 0)   V(−3, −8)

Give the ordered pair for each point on the coordinate plane shown below.

19. A

20. B

21. C

22. D

23. E

24. F

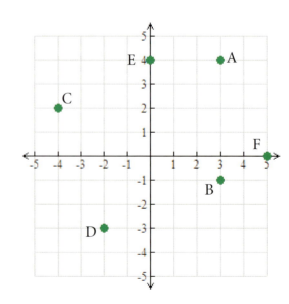

Determine the quadrant each point lies in or the axis the point lies on.

25. $(-6, 10)$

26. $(3, -1)$

27. $(0, -5)$

28. $(2, 9)$

29. $(-12, 0)$

30. $(-9, -9)$

Lesson 4.4 ~ Quadrilaterals on the Coordinate Plane

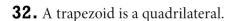

Determine if each statement is ALWAYS TRUE, SOMETIMES TRUE or NEVER TRUE.

31. A rhombus is a square.

32. A trapezoid is a quadrilateral.

33. A rectangle is a square.

34. A parallelogram is a trapezoid.

Graph each set of points on a coordinate plane. Connect the points in the order given. Connect the last point to the first. Name the shape with all terms that apply (parallelogram, rectangle, rhombus, square and/or trapezoid).

35. $(1, 4), (1, -2), (5, -2), (5, 4)$

36. $(-2, 3), (-1, -1), (4, -1), (3, 3)$

37. Find the perimeter and area of the shape formed by graphing the following points: $(-3, 2)$, $(-3, -1), (2, -1), (2, 2)$.

Lesson 4.5 ~ Input-Output Tables

38. Describe at least two ways you can display a mathematical relationship.

Copy and complete the input-output tables given the function rule.

39.

Input x	Function Rule $y = 7x$	Output y
0		
1		
2		
3		

40.

Input x	Function Rule $y = 3x - 2$	Output y
2		
4		
5		
8		

41. Write the four ordered pairs from the input-output table in **Exercise 40**.

42. Mikayla is able to walk 4 miles per hour. She wants to know how many miles (y) she can walk based on the number of hours (x) she has been walking. She created a function rule to help: $y = 4x$.

a. Copy and complete the table.

Input Hours, x	Function Rule $y = 4x$	Output Miles, y
1		
2		
3		
4		

b. Write the four ordered pairs from the table above by pairing the input and output values.

c. Draw a coordinate plane with an x-axis that goes from 0 to 10 and a y-axis that goes from 0 to 40 as seen below. Graph the ordered pairs on the coordinate plane.

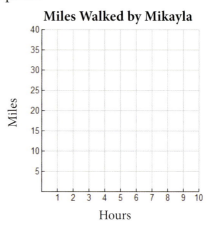

d. Connect the points with a line. This line shows Mikayla's continuous progress on her walk.

e. Mikayla entered an event named "Walk For A Cancer Cure". It took her 8 hours to complete the walk. How many miles did she walk? Use words and/or numbers to show how you determined your answer.

Lesson 4.6 ~ Writing Function Rules

• •

Find the function rule for each input-output table.

43.

Input x	Output y
0	14
1	11
2	8
3	5

44.

Input x	Output y
0	5
1	5.5
2	6
3	6.5

45. Use the function rule $y = 13 - 2x$ to answer the questions.

a. What is the amount of change? How do you know?

b. What is the start value?

c. Is the graph of this function increasing or decreasing? How do you know?

Find the function rule for each graph.

46.

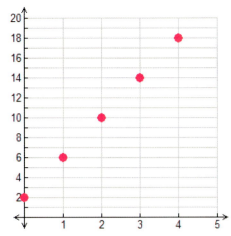

47.

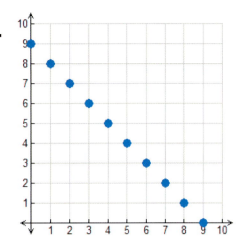

48. Andy's parents gave him $50 to open a savings account. Andy did a paper route to earn extra money. He earned enough to put $15 in his savings account each week.

    a. Copy and complete the input-output table based on the information above.

Input Weeks, x	Output Money In Account, y
0	
1	
2	
3	
4	
5	

    b. Find the function rule for the table.

    c. How much money will Andy have in his savings account after 9 weeks? Show all work necessary to justify your answer.

Lesson 4.7 ~ Graphing Linear Functions

Graph each linear function using input-output tables.

49. $y = 4 + 2x$ **50.** $y = 7 - x$ **51.** $y = 2.5x$

Graph each linear function using start values and amounts of change.

52. $y = 1 + 3x$ **53.** $y = 8 - 3x$ **54.** $y = 2 + x$

55. Markesha is training to run a marathon. She runs at a rate of 7 miles per hour. Let x represent the number of hours since Markesha began running. Let y represent the total number of miles she ran. Write a function rule that models this situation. Graph the function.

Copy each recursive sequence. Fill in the missing values. Identify the first term and the operation that must be performed to arrive at the next term.

56. 25, 21, 17, _____, _____, _____

57. 7, 15, 23, _____, _____, _____

58. 40, _____, 70, _____, _____

59. 20, _____, 16, 14, _____, _____

60. Cubes, with faces each measuring one square inch, are stacked on top of each other.

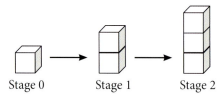

Stage 0   Stage 1   Stage 2

    a. What is the surface area of Stage 0 using just one cube?

    b. What is the surface area of the figure using two cubes? Three cubes?

    c. Draw the next stage in the pattern. What is the surface area of this figure?

    d. What is the operation that describes the surface area from one stage to the next?

    e. What will be the surface area of the figure formed in Stage 7? Explain how you determined your answer.

Write an inequality for each graph shown. Use *x* as the variable.

61.

62.

63.

64.

Write an inequality for each statement AND graph the inequality on a number line.

65. Susan runs at least 2 miles (*m*) each day.

66. Latesha ate less than 10 fries at lunch.

67. The temperature was greater than 45° outside.

68. Carl's kitten weighs 4 pounds at the most.

CAREER FOCUS

KASEY
VETERINARIAN

I am a veterinarian. I work with both large and small animals. I deal with all aspects of animal health, from dentistry to weight control. My job changes from day to day depending on what type of animal is brought in and what type of problems each is having. I give exams to make sure animals are healthy. I also help owners know how best to care for their animals. It is a very demanding job and I have to be ready to respond to emergencies 24 hours a day.

I use math multiple times a day. I use an animal's body weight in kilograms to calculate how much medicine to give them. Sometimes I am asked to let owners know how much of a certain feed to give their animals. I also give shots and need to know the quantity for each type of animal. A horse requires something very different than a house cat. I use math formulas and measurements to treat each one appropriately. Fractions and percentages are also types of math that play a big part in my day-to-day work.

To become a veterinarian requires getting a four year degree. Usually the degree will be in biology or animal science. After that, somebody planning to become a veterinarian has to go to a College of Veterinary Medicine. The total amount of schooling is about 8 years.

A veterinarian just out of school can expect to earn between $50,000 and $55,000 per year. Salaries increase according to which type of animal you work with and how much experience you have. Veterinarians who own a practice can earn over $100,000 per year.

Each day is different and challenging for a veterinarian. You never know what may walk through the door at any given time. I like the variety and challenges that come with veterinarian work. I also like interacting with animals and the personal growth I feel after learning something new.

ACKNOWLEDGEMENTS

All Photos and Clipart ©2008 Jupiterimages Corporation and Clipart.com with the exception of the cover photo and the following photos.

Introduction to Algebra Page 4
©iStockphoto.com/Alejandro Rivera

Introduction to Algebra Page 5
©iStockphoto.com/kieran wills

Introduction to Algebra Page 7
©iStockphoto.com/Darrin Henry

Introduction to Algebra Page 13
©iStockphoto.com/Focus_IN

Introduction to Algebra Page 16
©iStockphoto.com/Maridav

Introduction to Algebra Page 19
©iStockphoto.com/lawcain

Introduction to Algebra Page 19
©iStockphoto.com/1 design

Introduction to Algebra Page 19
©iStockphoto.com/Bennewitz

Introduction to Algebra Page 19
©iStockphoto.com/gülden özenli

Introduction to Algebra Page 19
©iStockphoto.com/Glenda Powers

Introduction to Algebra Page 20
©iStockphoto.com/dimdimich

Introduction to Algebra Page 21
©iStockphoto.com/mark wragg

Introduction to Algebra Page 23
©iStockphoto.com/james steidl

Introduction to Algebra Page 24
©iStockphoto.com/Juanmonino

Introduction to Algebra Page 31
©iStockphoto.com/Leslie Banks

Introduction to Algebra Page 32
©iStockphoto.com/PaulMcGuire

Introduction to Algebra Page 33
©iStockphoto.com/Gary Radler

Introduction to Algebra Page 34
©iStockphoto.com/Jacek Nowak

Introduction to Algebra Page 35
©iStockphoto.com/olaf herschbach

Introduction to Algebra Page 38
©iStockphoto.com/Grzegorz Kula

Introduction to Algebra Page 38
©iStockphoto.com/DNY59

Introduction to Algebra Page 41
©iStockphoto.com/Dmytro Sukharevskyy

Introduction to Algebra Page 41
©iStockphoto.com/Nicholas Rjabow

Introduction to Algebra Page 43
©iStockphoto.com/Aaron Kohr

Introduction to Algebra Page 44
©iStockphoto.com/Christopher Futcher

Introduction to Algebra Page 49
©iStockphoto.com/Julián Rovagnati

Introduction to Algebra Page 50
©iStockphoto.com/Victoria Chukalina

Introduction to Algebra Page 53
©iStockphoto.com/pink_cotton_candy

Introduction to Algebra Page 56
©iStockphoto.com/Neustockimages

Introduction to Algebra Page 59
©iStockphoto.com/Steve Debenport

Introduction to Algebra Page 60
©iStockphoto.com/Simone Becchetti

Introduction to Algebra Page 62
©iStockphoto.com/James Trice

Introduction to Algebra Page 63
©iStockphoto.com/Michel Aubry

Introduction to Algebra Page 63
©iStockphoto.com/igor terekhov

Introduction to Algebra Page 64
©iStockphoto.com/Eric Hood

Introduction to Algebra Page 71
©iStockphoto.com/MachineHeadz

Introduction to Algebra Page 72
©iStockphoto.com/Pavel Losevsky

Introduction to Algebra Page 76
©iStockphoto.com/Jon Patton

Introduction to Algebra Page 79
©iStockphoto.com/Lisa F. Young

Introduction to Algebra Page 79
©iStockphoto.com/Yvan Dubé

Introduction to Algebra Page 82
©iStockphoto.com/MachineHeadz

Introduction to Algebra Page 86
©iStockphoto.com/Jakub Jirsák

Introduction to Algebra Page 90
©iStockphoto.com/Žiga Četrtič

Introduction to Algebra Page 92
©iStockphoto.com/Jani Bryson

Introduction to Algebra Page 92
©iStockphoto.com/Nicholas Moore

Introduction to Algebra Page 93
©iStockphoto.com/Pictac

Introduction to Algebra Page 99
©iStockphoto.com/Barbro Bergfeldt

Introduction to Algebra Page 104
©iStockphoto.com/Studio-Annika

Introduction to Algebra Page 104
©iStockphoto.com/Chris Elwell

Introduction to Algebra Page 105
©iStockphoto.com/Maridav

Introduction to Algebra Page 106
©iStockphoto.com/FredFroese

Introduction to Algebra Page 112
©iStockphoto.com/kristian sekulic

Introduction to Algebra Page 112
©iStockphoto.com/Brian Hanger

Introduction to Algebra Page 116
©iStockphoto.com/maogg

Introduction to Algebra Page 116
©iStockphoto.com/mbbirdy

Introduction to Algebra Page 117
©iStockphoto.com/desifoto

Introduction to Algebra Page 118
©iStockphoto.com/Tamás Ambrits

Introduction to Algebra Page 119
©iStockphoto.com/kristian sekulic

Introduction to Algebra Page 120
©iStockphoto.com/abu

Introduction to Algebra Page 122
©iStockphoto.com/Natallia Khlapushyna

Introduction to Algebra Page 123
©iStockphoto.com/Rebecca Paul

Introduction to Algebra Page 124
©iStockphoto.com/Tomas Anderson

Introduction to Algebra Page 125
©iStockphoto.com/Mladen Mitrinović

Introduction to Algebra Page 128
©iStockphoto.com/Merih Unal Ozmen

Introduction to Algebra Page 135
©iStockphoto.com/diego cervo

Introduction to Algebra Page 137
©iStockphoto.com/poligonchik

Introduction to Algebra Page 138
©iStockphoto.com/Stockphoto4u

Introduction to Algebra Page 151
©iStockphoto.com/Ash Waechter

Introduction to Algebra Page 152
©iStockphoto.com/PhotoTalk

Introduction to Algebra Page 153
©iStockphoto.com/xyno6

Introduction to Algebra Page 155
©iStockphoto.com/Robert Dant

Introduction to Algebra Page 156
©iStockphoto.com/bora ucak

Layout and Design by Judy St. Lawrence

Cover Design by Schuyler St. Lawrence

Glossary Translation by Keyla Santiago and Heather Contreras

CORE FOCUS ON MATH
GLOSSARY ~ GLOSARIO

A

Absolute Value — The distance a number is from 0 on a number line.

Valor Absoluto — La distancia de un número desde el 0 en una recta numérica.

Acute Angle — An angle that measures more than 0° but less than 90°.

Ángulo Agudo — Un ángulo que mide mas 0° pero menos de 90°.

Adjacent Angles — Two angles that share a ray.

Ángulos Adyacentes — Dos ángulos que comparten un rayo.

Algebraic Expression — An expression that contains numbers, operations and variables.

Expresiones Algebraicas — Una expresión que contiene números, operaciones y variables.

Alternate Exterior Angles — Two angles that are on the outside of two lines and are on opposite sides of a transversal.

Ángulos Exteriores Alternos — Dos ángulos que están afuera de dos rectas y están a lados opuestos de una transversal.

Alternate Interior Angles — Two angles that are on the inside of two lines and are on opposites sides of a transversal.

Ángulos Interiores Alternos — Dos ángulos que están adentro de dos rectas y están a lados opuestos de una transversal.

Angle — A figure formed by two rays with a common endpoint.

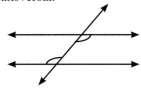

Ángulo — Una figura formada por dos rayos con un punto final en común.

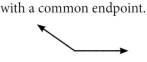
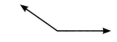

Area	The number of square units needed to cover a surface.	Área	El número de unidades cuadradas necesitadas para cubrir una superficie.
Ascending Order	Numbers arranged from least to greatest.	Progresión Ascendente	Los números ordenados de menor a mayor.
Associative Property	A property that states that numbers in addition or multiplication expressions can be grouped without affecting the value of the expression.	Propiedad Asociativa	Una propiedad que establece que los números en expresiones de suma o de multiplicación pueden ser agrupados sin afectar el valor de la expresión.
Axes	A horizontal and vertical number line on a coordinate plane.	Ejes	Una recta numérica horizontal y vertical en un plano de coordenadas.

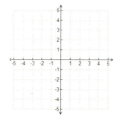 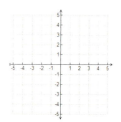

Axis of Symmetry	The line of symmetry on a parabola that goes through the vertex.	El Eje De Las Simetría	La linia de simetría de una parábola que pasa por el vértice.

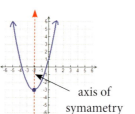

axis of symametry

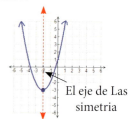

El eje de Las simetria

B

Bar Graph	A graph that uses bars to compare the quantities in a categorical data set.	Gráfico de Barras	Una gráfica que utiliza barras para comparar las cantidades en un conjunto de datos categórico.

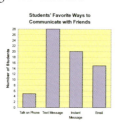 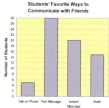

Base (of a power)	The base of the power is the repeated factor. In x^a, x is the base.	Base (de la potencia)	La base de la potenciación es el factor repatidio. En x^a, x es la base.

Base (of a solid)	See Prism, Cylinder, Pyramid and Cone.	Base (de un sólido)	Ver Prisma, Cilindro, Pirámide y Cono.
Base (of a triangle)	Any side of a triangle.	Base (de un triángulo)	Cualquier lado de un triángulo.
Bias	A problem when gathering data that affects the results of the data.	Sesgo	Un problema que ocurre cuando se recogen datos que afectan los resultados de los datos.
Biased Sample	A group from a population that does not accurately represent the entire population.	Muestra Sesgada	Un grupo de una población que no representa con exactitud la población entera.
Binomials	Expressions involving two terms (i.e. $x - 2$).	Binomiales	Expresiones que impliquen dos terminos. (es decir: $x - 2$).
Bivariate Data	Data that describes two variables and looks at the relationship between the two variables.	Datos de dos Variables	Los datos que describen dos variables y analiza la relación entre estas dos variables.
Box-and-Whisker Plot	A diagram used to display the five-number summary of a data set.	Diagrama de Líneas y Bloques	Un diagrama utilizado para mostrar el resumen de cinco números de un conjunto de datos.

C

Categorical Data	Data collected in the form of words.	Datos Categóricos	Datos recopilados en la forma de palabras.
Center of a Circle	The point inside a circle that is the same distance from all points on the circle.	Centro de un Círculo	Un ángulo dentro de un círculo que está a la misma distancia de todos los puntos en el círculo.

Central Angle	An angle in a circle with its vertex at the center of a circle.	Ángulo Central	Un ángulo en un círculo con su vértice en el centro del círculo.
Chord	A line segment with endpoints on a circle.	Cuerda	Un segmento de la recta con puntos finales en el círculo.
Circle	The set of all points that are the same distance from a center point.	Círculo	El conjunto de todos los puntos que están a la misma distancia de un punto central.
Circumference	The distance around a circle.	Circunferencia	La distancia alrededor de un círculo.
Coefficient	The number multiplied by a variable in a term.	Coeficiente	El número multiplicado por una variable en un término.
Commutative Property	A property that states numbers can be added or multiplied in any order.	Propiedad Conmutativa	Una propiedad que establece que los números pueden ser sumados o multiplicados en cualquier orden.
Compatible Numbers	Numbers that are easy to mentally compute; used when estimating products and quotients.	Números Compatibles	Números que son fáciles de calcular mentalmente; utilizado cuando se estiman productos y cocientes.
Complementary Angles	Two angles whose sum is 90°.	Ángulos Complementarios	Dos ángulos cuya suma es de 90°.
Complements	Two probabilities whose sum is 1. Together they make up all the possible outcomes without repeating any outcomes.	Complementos	Dos probabilidades cuya suma es de 1. Juntos crean todos los posibles resultados sin repetir alguno.

Completing the Square	The creation of a perfect square trinomial by adding a constant to an expression in the form $x^2 + bx$.	Terminado el Cuadrado	La creación de un trinomio cuadrado perfecto por adición de una constante a una expresión en la forma $x^2 + bx$.
Complex Fraction	A fraction that contains a fractional expression in its numerator, denominator or both. $$\frac{\frac{3}{4}}{\frac{3}{8}}$$	Fracción Compleja	Una fracción que contiene una expresión fraccionaria en su numerador, el denominador o ambos. $$\frac{\frac{3}{4}}{\frac{3}{8}}$$
Composite Figure	A geometric figure made of two or more geometric shapes.	Figura Compuesta	Una figura geométrica formada por dos o más formas geométricas.
Composite Number	A whole number larger than 1 that has more than two factors.	Número Compuesto	Un número entero mayor que el 1 con más de dos factores.
Composite Solid	A solid made of two or more three-dimensional geometric figures.	Sólido Compuesto	Un sólido formado por dos o más figuras geométricas tridimensionales.
Composition of Transformations	A series of transformations on a point.	Composición de Transformaciones	Una serie de transformaciones en un punto.
Compound Probability	The probability of two or more events occurring.	Compuesto de Probabilidad	La probabilidad de dos o más eventos ocurriendo.
Conditional Frequency	The ratio of the observed frequency to the total number of frequencies in a given category from an experiment or survey.	Frecuencia Condicional	La relación de una frecuencia observada para el número total de frecuencias en una categoría dada del experimento o encuesta.
Cone	A solid formed by one circular base and a vertex.	Cono	Un sólido formado por una base circular y una vértice.
Congruent	Equal in measure.	Congruente	Igual en medida.

Congruent Figures	Two shapes that have the exact same shape and the exact same size.	**Figuras Congruentes**	Dos figuras que tienen exactamente la misma forma y el mismo tamaño.

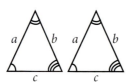
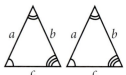

Constant	A term that has no variable.	**Constante**	Un término que no tiene variable.
Continuous	When a graph can be drawn from beginning to end without any breaks.	**Continuo**	Cuando una gráfica puede ser dibujada desde principio a fin sin ninguna interrupción.
Conversion	The process of renaming a measurement using different units.	**Conversión**	El proceso de renombrar una medida utilizando diferentes unidades.
Coordinate Plane	A plane created by two number lines intersecting at a 90° angle.	**Plano de Coordenadas**	Un plano creado por dos rectas numéricas que se intersecan a un ángulo de 90°.

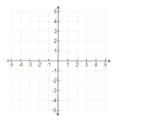
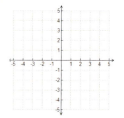

Correlation	The relationship between two variables in a scatter plot.	**Correlación**	La relación entre dos variables en un gráfico de dispersión.
Corresponding Angles	Two non-adjacent angles that are on the same side of a transversal with one angle inside the two lines and the other on the outside of the two lines.	**Ángulos Correspondientes**	Dos ángulos no adyacentes que están en el mismo lado de una transversal con un ángulo adentro de las dos rectas y el otro afuera de las dos rectas.

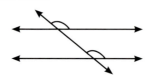

Corresponding Parts	The angles and sides in similar or congruent figures that match.	**Partes Correspondientes**	Los ángulos y lados en figuras similares o congruentes que concuerdan.

Cube Root	One of the three equal factors of a number. $$3 \cdot 3 \cdot 3 = 27 \qquad \sqrt[3]{27} = 3$$	Raíz Cúbica	Uno de los tres factores iguales de un número. $$3 \cdot 3 \cdot 3 = 27 \qquad \sqrt[3]{27} = 3$$
Cubed	A term raised to the power of 3.	Cubicado	Un término elevado a la potencia de 3.
Cylinder	A solid formed by two congruent and parallel circular bases.	Cilindro	Un sólido formado por dos bases circulares congruentes y paralelas.

D

Decimal	A number with a digit in the tenths place, hundredths place, etc.	Decimal	Un número con un dígito en las décimas, las centenas, etc.
Degrees	A unit used to measure angles.	Grados	Una unidad utilizada para medir ángulos.
Dependent Events	Two (or more) events such that the outcome of one event affects the outcome of the other event(s).	Eventos Dependiente	Dos (o más) eventos de tal manera que el resultado de un evento afecta el resultado del otro evento (s).
Dependent Variable	The variable in a relationship that depends on the value of the independent variable.	Variable Dependiente	La variable en una relación que depende del valor de la variable independiente.
Descending Order	Numbers arranged from greatest to least.	Progresión Descendente	Los números ordenados de mayor a menor.

Diameter	The distance across a circle through the center.	Diámetro	La distancia a través de un círculo por el centro.
Dilation	A transformation which changes the size of the figure but not the shape. 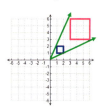	Dilatación	Una transformación que cambia el tamaño de la figura, pero no la forma. 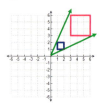
Direct Variation	A linear function that passes through the origin and has the equation $y = mx$.	Variación Directa	Una función lineal que pasa a través del origen y tiene la ecuación $y = mx$.
Discount	The decrease in the price of an item.	Descuento	La disminución de precio en un artículo.
Discrete	When a graph can be represented by a unique set of points rather than a continuous line.	Discreto	Cuando una gráfica puede ser representada por un conjunto de puntos único en vez de una recta continua.
Discriminant	In the quadratic formula, the expression under the radical sign. The discriminant provides information about the number of real roots or solutions of a quadratic equation. $$\frac{-b \pm \sqrt{b^2 - 4ac}}{2a}$$	Discriminante	En la fórmula cuadrática, la expresión bajo el signo radical. El discriminante proporciona información sobre el número o las verdaderas raíces o soluciones de una ecuación cuadrática. $$\frac{-b \pm \sqrt{b^2 - 4ac}}{2a}$$
Distance Formula	A formula used to find the distance between two points on the coordinate plane. $$d = \sqrt{(x_2 - x_1)^2 + (y_2 - y_1)^2}$$	Fórmula de Distancia	Una fórmula utilizada para encontrar la distancia entre dos puntos en un plano de coordenadas. $$d = \sqrt{(x_2 - x_1)^2 + (y_2 - y_1)^2}$$

Distributive Property	A property that can be used to rewrite an expression without parentheses. $a(b + c) = ab + ac$	**Propiedad Distributiva**	Una propiedad que puede ser utilizada para reescribir una expresión sin paréntesis: $a(b + c) = ab + ac$
Dividend	The number being divided. **100** $\div$ 4 = 25	**Dividendo**	El número que es dividido. **100** $\div$ 4 = 25
Divisor	The number used to divide. 100 $\div$ **4** = 25	**Divisor**	El número utilizado para dividir. 100 $\div$ **4** = 25
Domain	The set of input values of a function.	**El Dominio**	El conjunto de valores entrados de la función.
Dot Plot	A data display that consists of a number line with dots equally spaced above data values. 	**Punto de Gráfico**	Una visualización de datos que consiste de una línea numérica con puntos igualmente espaciados sobre valores de datos. 
Double Stem-and-Leaf Plot	A stem-and-leaf plot where one set of data is placed on the right side of the stem and another is placed on the left of the stem. 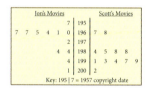	**Doble Gráfica de Tallo y Hoja**	Una gráfica de tallo y hoja donde un conjunto de datos es colocado al lado derecho del tallo y el otro es colocado a la izquierda del tallo. 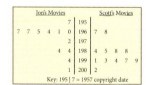

E

Edge	The segment where two faces of a solid meet. 	**Arista (Borde)**	El segmento donde dos caras de un sólido se encuentran.

Elimination Method	A method for solving a system of linear equations.	Método de Eliminación	Un método para resolver un sistema de ecuaciones lineales.
Enlargement	A dilation that creates an image larger than its pre-image.	Agrandamiento	Una dilatación que crea una imagen más grande que su pre-imagen.
Equally Likely	Two or more possible outcomes of a given situation that have the same probability.	Igualmente Probables	Dos o más posibles resultados de una situación dada que tienen la misma probabilidad.
Equation	A mathematical sentence that contains an equals sign between 2 expressions.	Ecuación	Una oración matemática que contiene un símbolo de igualdad entre dos expresiones.
Equiangular	A polygon in which all angles are congruent.	Equiángulo	Un polígono en el cual todos los ángulos son congruentes.
Equilateral	A polygon in which all sides are congruent.	Equilátero	Un polígono en el cual todos los lados son congruentes.

Equivalent Decimals	Two or more decimals that represent the same number.	Decimales Equivalentes	Dos o más decimales que representan el mismo número.
Equivalent Expressions	Two or more expressions that represent the same algebraic expression.	Expresiones Equivalentes	Dos o más expresiones que representan la misma expresión algebraica.
Equivalent Fractions	Two or more fractions that represent the same number.	Fracciones Equivalentes	Dos o más fracciones que representan el mismo número.
Evaluate	To find the value of an expression.	Evaluar	Encontrar el valor de una expresión.
Even Distribution	A set of data values that is evenly spread across the range of the data.	Distribución Igualada	Un conjunto de valores de datos que es esparcido de modo uniforme a través del rango de los datos.

Event	A desired outcome or group of outcomes.	Evento	Un resultado o grupo de resultados deseados.
Experimental Probability	The ratio of the number of times an event occurs to the total number of trials.	Probabilidad Experimental	La razón de la cantidad de veces que un suceso ocurre a la cantidad total de intentos.
Exponent	In x^a, a is the exponent. The exponent shows the number of times the factor (x) is repeated.	Exponente	En x^a, a es el exponente. El exponente indica el número de veces que se repite el factor (x).
Exponential Function	A function that can be described by an equation in the form $f(x) = bm^x$.	Función Exponencial	Una función que puede ser descrito por una ecuación en la forma $f(x) = bm^x$.

F

Face	A polygon that is a side or base of a solid.	Cara	Un polígono que es una base de lado de un sólido.

Factors	Whole numbers that can be multiplied together to find a product.	Factores	Números enteros que pueden ser multiplicados entre si para encontrar un producto.
First Quartile (Q1)	The median of the lower half of a data set.	Primer Cuartil (Q1)	Mediana de la parte inferior de un conjunto de datos.
Five-Number Summary	Describes the spread of a data set using the minimum, 1st quartile, median, 3rd quartile and maximum.	Sumario de Cinco Números	Describe la extensión de un conjunto de datos utilizando el mínimo, el primer cuartil, la mediana el tercer cuartil y el máximo.
Formula	An algebraic equation that shows the relationship among specific quantities.	Fórmula	Una ecuación algebraica que enseña la relación entre cantidades específicas.
Fraction	A number that represents a part of a whole number, written as $\frac{numerator}{denominator}$.	Fracción	Un número que representa una parte de un número entero, escrito como $\frac{numerador}{denominador}$.

| Frequency | The number of times an item occurs in a data set. | Frecuencia | La cantidad de veces que un artículo ocurre en un conjunto de datos. |

| Frequency Table | A table which shows how many times a value occurs in a given interval. | Tabla de Frecuencia | Una tabla que enseña cuantas veces un valor ocurre en un intervalo dado. |

Weight of Newborn (in Pounds)	Tally
4 – 5.5	|
5.5 – 7	|||
7 – 8.5	ⅢⅢ
8.5 – 10	||
10 – 11.5	|

| Function | A relationship between two variables that has one output value for each input value. | Función | Una relación entre dos variables que tiene un valor de salida para cada valor de entrada. |

G

| General Form | A quadratic function is in general form when written $f(x) = ax^2 + bx + c$ where $a \neq 0$. | Forma General | Una función cuadrática es en forma general cuándo escrito $f(x) = ax^2 + bx + c$ donde $a \neq 0$. |

| Geometric Probability | Ratios of lengths or areas used to find the likelihood of an event. | Probabilidad Geométrica | Razones de longitudes o áreas utilizadas para encontrar la probabilidad de un suceso. |

| Geometric Sequence | A list of numbers that begins with a starting value. Each term in the sequence is generated by multiplying the previous term in the sequence by a constant multiplier. | Secuenciación Geométrica | Una lista de números que comienza con un valor inicial. Cada término de la secuencia se genera al multiplicar el término anterior de la secuencia por un multiplicar constante. |

| Greatest Common Factor (GCF) | The greatest factor that is common to two or more numbers. | Máximo Común Divisor (MCD) | El máximo divisor que le es común a dos o más números. |

| Grouping Symbols | Symbols such as parentheses or fraction bars that group parts of an expression. | Símbolos de Agrupación | Símbolos como el paréntesis o barras de fracción que agrupan las partes de una expresión. |

H

Height of a Triangle	A perpendicular line drawn from the side whose length is the base to the opposite vertex.	**Altura de un Triángulo**	Una recta perpendicular dibujada desde el lado cuya longitud es la base al vértice opuesto.
Histogram	A bar graph that displays the frequency of numerical data in equal-sized intervals. 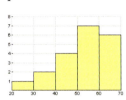	**Histograma**	Un gráfico de barras que muestra la frecuencia de datos numéricos en intervalos de tamaños iguales. 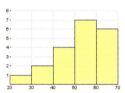
Hypotenuse	The side opposite the right angle in a right triangle. 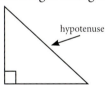	**Hipotenusa**	El lado opuesto el ángulo recto en un triángulo rectángulo. 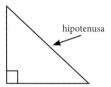

I-J-K

Image	A point or figure which is the result of a transformation or series of transformations.	**Imagen**	Un punto o figura que es el resultado de una transformación o una serie de transformaciones.
Improper Fraction	A fraction whose numerator is greater than or equal to its denominator.	**Fracción Impropia**	Una fracción cuyo numerador es mayor o igual a su denominador.
Independent Events	Two (or more) events such that the outcome of one event does not affect the outcome of the other event(s).	**Eventos Independientes**	Dos (o más) eventos de tal manera que el resultado de un evento no afecta el resultado del otro evento (s).

English	Definition	Spanish	Definición
Independent Variable	The variable representing the input values. ←independent	**Variable Independiente**	La variable que representa los valores entratos. ←independiente
Inequality	A mathematical sentence using $<$, $>$, $\leq$ or $\geq$ to compare two quantities.	**Desigualdad**	Un enunciado matemático usando $<$, $>$, $\leq$ ó $\geq$ para comparar dos cantidades.
Inference	A logical conclusion based on known information.	**Inferencia**	Una conclusión lógica basada en la información conocida.
Input-Output Table	A table used to describe a function by listing input values with their output values.	**Tabla de Entrada y Salida**	Una tabla utilizada para describir una función al enumerar valores de entrada con sus valores de salidas.
Integers	The set of all whole numbers, their opposites, and 0.	**Enteros**	El conjunto de todos los números enteros, sus opuestos y 0.
Interquartile Range (IQR)	The difference between the 3rd quartile and the 1st quartile in a set of data.	**Rango Intercuartil (IQR)**	La diferencia entre el tercer cuartil y el primer cuartil en un conjunto de datos.
Inverse Operations	Operations that undo each other.	**Operaciones Inversas**	Operaciones que se cancelan la una a la otra.
IQR Method	A method for determining outliers using interquartile ranges.	**Método IQR**	Un método para determinar los datos aberrantes.
Irrational Numbers	A number that cannot be expressed as a fraction of two integers.	**Números Irracionales**	Un número que no puede ser expresado como una fracción de dos enteros.

Isosceles Trapezoid	A trapezoid that has congruent legs.	Trapezoide Isósceles	Un trapezoide con catetos congruentes.

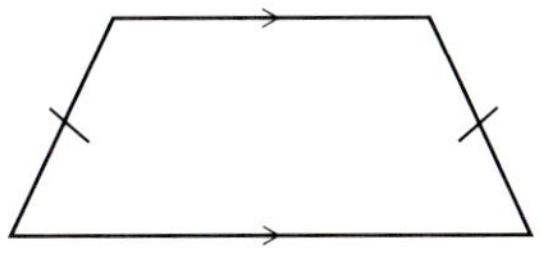

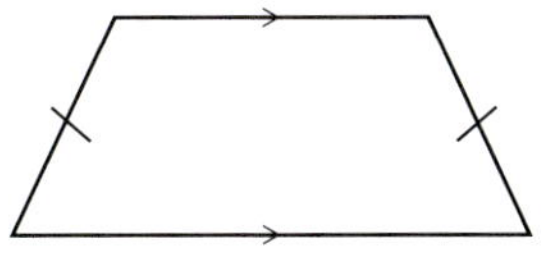

Isosceles Triangle	A triangle that has two or more congruent sides.	Triángulo Isósceles	Un triángulo que tiene dos o más lados congruentes.

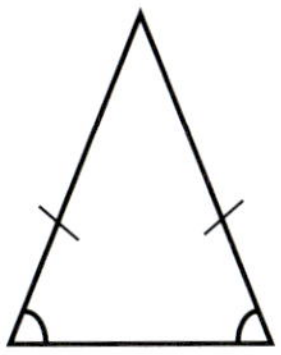

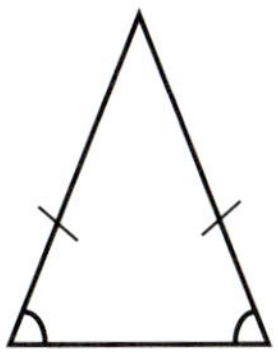

L

Lateral Face	A side of a solid that is not a base.	Cara Lateral	Un lado de un sólido que no sea una base.
Least Common Denominator (LCD)	The least common multiple of two or more denominators.	Mínimo Común Denominador (MCD)	El mínimo común múltiplo de dos o más denominadores.
Least Common Multiple (LCM)	The smallest nonzero multiple that is common to two or more numbers.	Mínimo Común Múltiplo (MCM)	El múltiplo más pequeño que no sea cero que le es común a dos o más números.
Leg	The two sides of a right triangle that form a right angle.	Cateto	Los dos lados de un triángulo rectángulo que forman un ángulo recto.

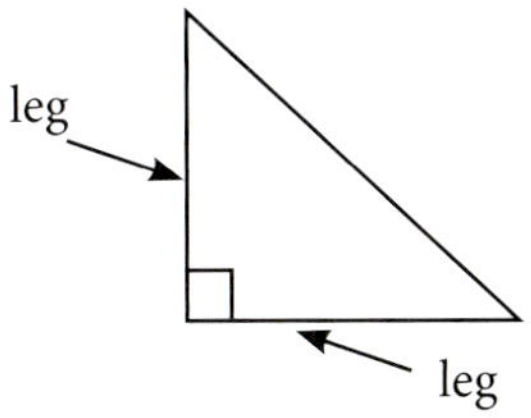

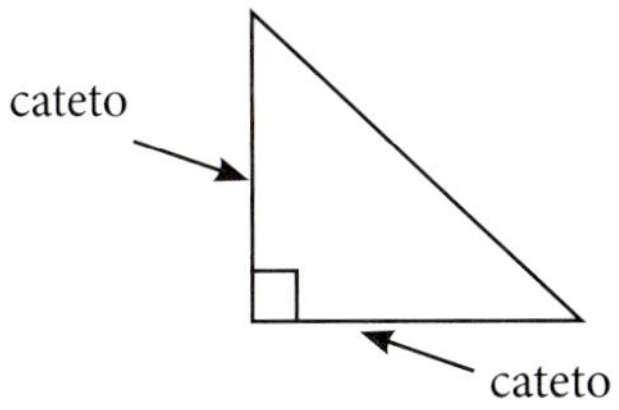

leg

leg

cateto

cateto

Like Terms	Terms that have the same variable(s).	Términos Semejantes	Términos que tienen el mismo variable(s).

| Line of Best Fit | A line which best represents the pattern of a two-variable data set. | Recta de Mejor Ajuste | Una recta que mejor representa el patrón de un conjunto de datos de dos variables. |

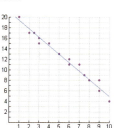 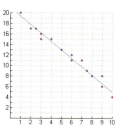

| Linear Equation | An equation whose graph is a line. | Ecuación Lineal | Una ecuación cuya gráfica es una recta. |

| Linear Function | A function whose graph is a line. | Función Lineal | Una función cuya gráfica es una recta. |

| Linear Pair | Two adjacent angles whose non-common sides are opposite rays. | Par Lineal | Dos ángulos adyacentes cuyos lados no comunes son rayos opuestos. |

M

| Mark-up | The increase in the price of an item. | Margen de Beneficio | El aumento de precio en un artículo. |

| Maximum | The highest point on a curve. | Máximo | El punto más alto en la curva. |

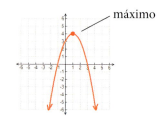

| Mean | The sum of all values in a data set divided by the number of values. | Media | La suma de todos los valores en un conjunto de datos dividido entre la cantidad de valores. |

| Mean Absolute Deviation | A statistic that shows the average distance from the mean for all numbers in a data set. | Desviación Media Absoluta | Una estadística que muestra la distancia promedio entre la media de todos los números en una serie de datos. |

Measures of Center	Numbers that are used to represent a data set with a single value; the mean, median, and mode are the measures of center.	Medidas de Centro	Números que son utilizados para representar un conjunto de datos con un solo valor; la media, la mediana, y la moda son las medidas de centro.
Measures of Variability	Statistics that help determine the spread of numbers in a data set.	Medidas de Variabilidad	Las estadísticas que ayudan a determinar la extensión de los números en una serie de datos.
Median	The middle number or the average of the two middle numbers in an ordered data set.	Mediana	El número medio o el promedio de los dos números medios en un conjunto de datos ordenados.
Minimum	The lowest point on a curve.	Mínimo	El punto más bajo en la curva.

Mixed Number	The sum of a whole number and a fraction less than 1.	Números Mixtos	La suma de un número entero y una fracción menor que 1.
Mode	The number(s) or item(s) that occur most often in a data set.	Moda	El número(s) o artículo(s) que ocurre con más frecuencia en un conjunto de datos.
Motion Rate	A rate that compares distance to time.	Índice de Movimiento	Un índice que compara distancia por tiempo.
Multiple	The product of a number and nonzero whole number.	Múltiplo	El producto de un número y un número entero que no sea cero.

N

Negative Number	A number less than 0.	Número Negativo	Un número menor que 0.

Net	A two-dimensional pattern that folds to form a solid.	Red	Un patrón bidimensional que se dobla para formar un sólido.

Non-Linear Function	A function whose graph does not form a line.	Ecuación No Lineal	Una ecuación cuya gráfica no forma una recta.
Normal Distribution	A set of data values where the majority of the values are located in the middle of the data set and can be displayed by a bell-shaped curve.	Distribución Normal	Un conjunto de valores de datos donde la mayoría de los valores están localizados en el medio del conjunto de datos y pueden ser mostrados por una curva de forma de campana.
Numerical Data	Data collected in the form of numbers.	Datos Numéricos	Datos recopilados en la forma de números.
Numerical Expressions	An expression consisting of numbers and operations that represents a specific value.	Expresiones Numéricas	Una expresión que consta de números y operaciones que representa un valor específico.

O

Obtuse Angle	An angle that measures more than 90° but less than 180°.	Ángulo Obtuso	Un ángulo que mide más de 90° pero menos de 180°.
Opposites	Numbers that are the same distance from 0 on a number line but are on opposite sides of 0.	Opuestos	Números a la misma distancia del 0 en un recta numérica pero en lados opuestos del 0.
Order of Operations	The rules to follow when evaluating an expression with more than one operation.	Orden de Operaciones	Las reglas a seguir cuando se evalúa una expresión con más de una operación.
Ordered Pair	A pair of numbers used to locate a point on a coordinate plane (x, y).	Par Ordenado	Un par de números utilizados para localizar un punto en un plano de coordenadas (x, y).

| Origin | The point where the *x*- and *y*-axis intersect on a coordinate plane (0, 0). | Origen | El punto donde el eje de la *x*-*y* el de la *y*- se cruzan en un plano de coordenadas (0,0). |

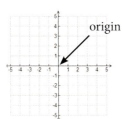 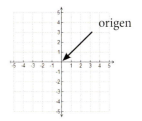

| Outcome | One possible result from an experiment or a probability sample space. | Resultado | Un resultado posible de un experimento o un espacio de probabilidad de la muestra. |

| Outlier | An extreme value that varies greatly from the other values in a data set. | Dato Aberrante | Un valor extremo que varía mucho de los otros valores en un conjunto de datos. |

P

| Parabola | The graph of a quadratic function. | Parábola | La gráfica de una función cuadratica. |

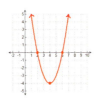

| Parallel | Lines in the same plane that never intersect. | Paralela | Rectas en el mismo plano que nunca se intersecan. |

| Parallel Box-and-Whisker Plot | One box-and-whisker plot placed above another; often used to compare data sets. | Diagrama Paralelo de Líneas y Bloques | Un diagrama de líneas y bloques ubicado sobre otro para comparar conjuntos de datos. |

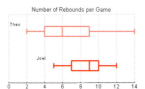 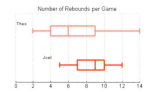

Parallelogram	A quadrilateral with both pairs of opposite sides parallel.	Paralelogramo	Un cuadrilateral con ambos pares de lados opuestos paralelos.

Parent Function	The simplest form of a particular type of function.	Función Principal	La forma más simple de un tipo particular de la función.
Parent Graph	The most basic graph of a function.	Gráfico Matriz	La gráfica más básica de una función.
Percent	A ratio that compares a number to 100.	Por Ciento	Una razón que compara un número con 100.
Percent of Change	The percent a quantity increases or decreases compared to the original amount.	Por Ciento de Cambio	El por ciento que una cantidad aumenta o disminuye comparado a la cantidad original.
Percent of Decrease	The percent of change when the new amount is less than the original amount.	Por Ciento de Disminución	El por ciento de cambio cuando la nueva cantidad es menos que la cantidad original.
Percent of Increase	The percent of change when the new amount is more than the original amount.	Por Ciento de Incremento	El por ciento de cambio cuando la nueva cantidad es más que la cantidad original.
Perfect Cube	A number whose cube root is an integer.	Cubo Perfecto	Un número cuyo raíz cúbica es un número entero.
Perfect Square	A number whose square root is an integer.	Cuadrado Perfecto	Un número cuyo raíz cuadrado es un número entero.
Perfect Square Trinomial	A trinomial that is the square of a binomial.	Trinomio Cuadrado Perfecto	Un trinomio que es el cuadrado de un binomio.
Perimeter	The distance around a figure.	Perímetro	La distancia alrededor de una figura.

Perpendicular	Two lines or segments that form a right angle.	**Perpendicular**	Dos rectas o segmentos que forman un ángulo recto.

Pi (π)	The ratio of the circumference of a circle to its diameter.	**Pi (π)**	La razón de la circunferencia de un círculo a su diámetro.
Pictograph	A graph that uses pictures to compare the amounts represented in a categorical data set.	**Gráfica Pictórica**	Una gráfica que utiliza dibujos para comparar las cantidades representadas en un conjunto de datos categóricos.

Pie Chart	A circle graph that shows information as sectors of a circle.	**Gráfico Circular**	Enseña la información como sectores de un círculo.

Polygon	A closed figure formed by three or more line segments.	**Polígono**	Una figura cerrada formada por tres o más segmentos de rectas.
Population	The entire group of people or objects one wants to gather information about.	**Población**	Todo el grupo de personas o los objetos a los que se quiere obtener información sobre.
Positive Number	A number greater than 0.	**Número Positivo**	Un número mayor que 0.
Power	An expression such as x^a which consists of two parts, the base (x) and the exponent (a).	**Potencia**	Una expresión como x^a que consiste de dos partes, la base (x) y el exponente (a).
Pre-image	The original figure prior to a transformation.	**Pre-imagen**	La figura original antes de una transformación.

Prime Factorization	When any composite number is written as the product of all its prime factors.	Factorización Prima	Cuando cualquier número compuesto es escrito como el producto de todos los factores primos.
Prime Number	A whole number larger than 1 that has only two possible factors, 1 and itself.	Número Primo	Un número entero mayor que 1 que tiene solo dos factores posibles, 1 y el mismo.
Prism	A solid formed by polygons with two congruent, parallel bases.	Prisma	Un sólido formado por polígonos con dos bases congruentes y paralelas.
Probability	The measure of how likely it is an event will occur.	Probabilidad	La medida de cuán probable un suceso puede ocurrir.
Product	The answer to a multiplication problem.	Producto	La respuesta a un problema de multiplicación.
Proper Fraction	A fraction with a numerator that is less than the denominator.	Fracción Propia	Una fracción con un numerador que es menos que el denominador.
Proportion	An equation stating two ratios are equivalent.	Proporción	Una ecuación que establece que dos razones son equivalentes.
Protractor	A tool used to measure angles.	Transportador	Una herramienta para medir ángulos.
Pyramid	A solid with a polygonal base and triangular sides that meet at a vertex.	Pirámide	Un sólido con una base poligonal y lados triangulares que se encuentran en un vértice.

| Pythagorean Triple | A set of three positive integers (a, b, c) such that $a^2 + b^2 = c^2$. | Triple de Pitágoras | Un conjunto de tres enteros positivos (a, b, c) tal que $a^2 + b^2 = c^2$. |

Q

| Q-Points | Points that are created by the intersection of the quartiles for the x- and y-values of a two-variable data set. | Puntos Q | Puntos que son creados por la intersección de los cuartiles para los valores de la x- y la y- de un conjunto de datos de dos variables. |

| Quadrants | Four regions formed by the x and y axes on a coordinate plane. | Cuadrantes | Cuatro regiones formadas por el eje-x y el eje-y en un plano de coordenadas. |

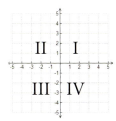 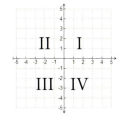

| Quadratic Formula | A method which can be used to solve quadratic equations in the form $0 = ax^2 + bx + c$, where $a \neq 0$. | Fórmula Cuadrática | Un métado que puede usarse para resolver ecuaciones cuadraticas en la forma $0 = ax^2 + bx + c$, donde $a \neq 0$. |

$$x = \frac{-b \pm \sqrt{b^2 - 4ac}}{2a}$$

$$x = \frac{-b \pm \sqrt{b^2 - 4ac}}{2a}$$

| Quadratic Function | Any function in the family with the parent function of $f(x) = x^2$. | Función Cuadrática | Cualquier otra función en la familia con la función principal de $f(x) = x^2$. |

| Quadrilateral | A polygon with four sides. | Cuadrilateral | Un polígono con cuatro lados. |

| Quotient | The answer to a division problem. | Cociente | La solución a un problema de división. |

R

| Radius | The distance from the center of a circle to any point on the circle. | Radio | La distancia desde el centro de un círculo a cualquier punto en el círculo. |

Random Sample	A group from a population created when each member of the population is equally likely to be chosen.	Muestra Aleatoria	Un grupo de una población creada cuando cada miembro de la población tiene la misma probabilidad de ser elegido.
Range (of a data set)	The difference between the maximum and minimum values in a data set.	Rango	La diferencia entre los valores máximo y mínimo en un conjunto de datos.
Range (of a function)	The set of output values of a function.	Rango (de una función)	El conjunto de valores salidos de la función.
Rate	A ratio of two numbers that have different units.	Índice	Una proporción de dos números con diferentes unidades.
Rate Conversion	A process of changing at least one unit of measurement in a rate to a different unit of measurement.	Conversión de Índice	Un proceso de cambiar por lo menos una unidad de medición en un índice a una diferente unidad de medición.
Rate of Change	The change in y-values over the change in x-values on a linear graph.	Índice de Cambio	El cambio en los valores de y sobre el cambio en los valores de x en una gráfica lineal.
Ratio	A comparison of two numbers using division. $a:b \quad \frac{a}{b} \quad a$ to b	Razón	Una comparación de dos números utilizando división. $a:b \quad \frac{a}{b} \quad a$ a b
Rational Number	A number that can be expressed as a fraction of two integers.	Número Racional	Un número que puede ser expresado como una fracción de dos enteros.
Ray	A part of a line that has one endpoint and extends forever in one direction.	Rayo	Una parte de una recta que tiene un punto final y se extiende eternamente en una dirección.
Real Numbers	The set of numbers that includes all rational and irrational numbers.	Números Racionales	El conjunto de números que incluye todos los números racionales e irracionales.

Reciprocals	Two numbers whose product is 1.	Recíprocos	Dos números cuyo producto es 1.
Rectangle	A quadrilateral with four right angles.	Rectángulo	Un cuadrilátero con cuatro ángulos rectos.

Recursive Routine	A routine described by stating the start value and the operation performed to get the following terms.	Rutina Recursiva	Una rutina descrita al exponer el valor del comienzo y la operación realizada para conseguir los términos siguientes.
Recursive Sequence	An ordered list of numbers created by a first term and a repeated operation.	Secuencia Recursiva	Una lista de números ordenados creada por un primer término y una operación repetida.
Reduction	A dilation that creates an image smaller than its pre-image.	Reducción	Una dilatación que crea una imagen más pequeña que su pre-imagen.
Reflection	A transformation in which a mirror image is produced by flipping a figure over a line.	Reflexión	Una transformación en el que se produce una imagen reflejada volteando una figura sobre una línea.

Relative Frequency	The ratio of the observed frequency to the total number of frequencies in an experiment or survey.	Frecuencia Relativa	La proporción de la frecuencia observada para el número total de frecuencias en un experimento o estudio.
Remainder	A number that is left over when a division problem is completed.	Remanente	Un número que queda cuando un problema de división se ha completado.
Repeating Decimal	A decimal that has one or more digits that repeat forever.	Decimal Repetitivo	Un decimal que tiene uno o más dígitos que se repiten eternamente.

Representative Sample	A group from a population that accurately represents the entire population.	**Muestra Representativa**	Un grupo de una población que representa con precisión toda la población.
Rhombus	A quadrilateral with four sides equal in measure.	**Rombo**	Un cuadrilátero con cuatro lados iguales en la medida.

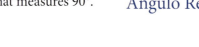

Right Angle	An angle that measures 90°.	**Ángulo Recto**	Un ángulo que mide 90°.

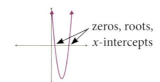 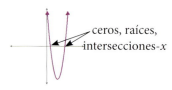

Roots — The *x*-intercepts of a quadratic function.

Raíces — Las intersecciones-*x* de una función cuadratica.

zeros, roots, *x*-intercepts

ceros, raíces, intersecciones-*x*

Rotation — A transformation which turns a point or figure about a fixed point, often the origin.

Rotación — Una transformación que convierte un punto a una figura sobre un punto fijo.

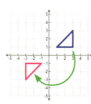 

S

Sales Tax — An amount added to the cost of an item. The amount added is a percent of the original amount as determined by a state, county or city.

Impuesto sobre las Ventas — Una cantidad añadida al costo de un artículo. La cantidad añadida es un por ciento de la cantidad original determinado por el estado, condado o ciudad.

Same-Side Interior Angles	Two angles that are on the inside of two lines and are on the same side of a transversal.	Ángulos Interiores del Mismo Lado	Dos ángulos que están en el interior de dos rectas y están en el mismo lado de una transversal.

 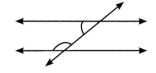

Sample	A group from a population that is used to make conclusions about the entire population.	Muestra	Un grupo de una población que se utiliza para sacar conclusiones sobre toda la población.

Sample Space	The set of all possible outcomes.	Muestra de Espacio	El conjunto de todos los resultados posibles.

Scale	The ratio of a length on a map or model to the actual object.	Escala	La razón de una longitud en un mapa o modelo al objeto verdadero.

Scale Factor	The ratio of corresponding sides in two similar figures.	Factor de Escala	La razón de los lados correspondientes en dos figuras similares.

Scalene Triangle	A triangle that has no congruent sides.	Triángulo Escaleno	Un triángulo sin lados congruentes.

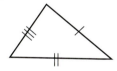 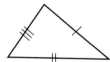

Scatter Plot	A set of ordered pairs graphed on a coordinate plane.	Diagrama de Dispersión	Un conjunto de pares ordenados graficados en un plano de coordenadas.

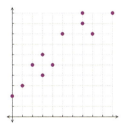 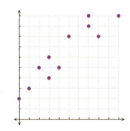

Scientific Notation	Scientific notation is an exponential expression using a power of 10 where $1 \leq N < 10$ and P is an integer. $N \times 10^P$	La Notación Científica	Notación científica es una expresión exponencial con una potencia de 10, donde $1 \leq N < 10$ y P es un número entero. $N \times 10^P$

Sector	A portion of a circle enclosed by two radii.	**Sector**	Una porción de un circulo encerado por dos radios.

Sequence	An ordered list of numbers.	**Sucesión**	Una lista de números ordenados.

Similar Figures	Two figures that have the exact same shape, but not necessarily the exact same size.	**Figuras Similares**	Dos figuras que tienen exactamente la misma forma, pero no necesariamente el mismo tamaño exacto.

Similar Solids	Solids that have the same shape and all corresponding dimensions are proportional.	**Sólidos Similares**	Sólidos con la misma forma y todas sus dimensiones correspondientes son proporcionales.

Simplest Form	A fraction whose numerator and denominator's only common factor is 1.	**Expresión Simple**	Una fracción cuyo único factor común del numerador y del denominador es 1.

Simplify an Expression	To rewrite an expression without parentheses and combine all like terms.	**Simplificar una Expresión**	Reescribir una expresión sin paréntesis y combinar todos los términos iguales.

Simulation	An experiment used to model a situation.	**Simulación**	Un experimento utilizado para modelar una situación.

Single-Variable Data	A data set with only one type of data.	**Datos de una Variable**	Un conjunto de datos con tan solo un tipo de datos.

Sketch	To make a figure free hand without the use of measurement tools.	**Esbozo**	Hacer una figura a mano libre sin utilizar herramientas de medidas.

Skewed Left	A plot or graph with a longer tail on the left-hand side.	**Torcido a la Izquierda**	Un gráfico con una cola al lado izquierdo.

Skewed Right	A plot or graph with a longer tail on the right-hand side.	**Torcido a la Derecha**	Un gráfico con una cola al lado derecho.
Slant Height	The height of a lateral face of a pyramid or cone.	**Altura Sesgada**	La altura de un cara lateral de una pirámide o cono.

Slope	The ratio of the vertical change to the horizontal change in a linear graph.	**Pendiente**	La razón del cambio vertical al cambio horizontal en una gráfica lineal.
Slope Triangle	A right triangle formed where one leg represents the vertical rise and the other leg is the horizontal run in a linear graph.	**Triángulo de Pendiente**	Un triángulo rectángulo formado donde una cateto representa el ascenso y la otra es una carrera horizontal en una gráfica lineal.

Slope-Intercept Form	A linear equation written in the form $y = mx + b$.	**Forma de las Intersecciones con la Pendiente**	Una ecuación lineal escrita en la forma $y = mx + b$.
Solid	A three-dimensional figure that encloses a part of space.	**Sólido**	Una figura tridimensional que encierra una parte del espacio.
Solution	Any value or values that makes an equation true.	**Solución**	Cualquier valor o valores que hacen una ecuación verdadera.
Solution of a System of Linear Equations	The ordered pair that satisfies both linear equations in the system.	**Solución de un Sistema de Ecuaciones Lineales**	El par ordenado que satisface ambas ecuaciones lineales en el sistema.

| Sphere | A solid formed by a set of points in space that are the same distance from a center point. | Esfera | Un sólido formado por un conjunto de puntos en el espacio que están a la misma distancia de un punto central.  |

Square — A quadrilateral with four right angles and four congruent sides.

Cuadrado — Un cuadrilátero con cuatro ángulos rectos y cuatro lados congruente.

Square Root — One of the two equal factors of a number.

$$3 \cdot 3 = 9 \qquad 3 = \sqrt{9}$$

Raíz Cuadrada — Uno de los dos factores iguales de un número.

$$3 \cdot 3 = 9 \qquad 3 = \sqrt{9}$$

Squared — A term raised to the power of 2.

Cuadrado — Un término elevado a la potencia de 2.

Start Value — The output value that is paired with an input value of 0 in an input-output table.

Valor de Comienzo — El valor de salida que es aparejado con un valor de entrada de 0 en una tabla de entradas y salidas.

Statistics — The process of collecting, displaying and analyzing a set of data.

Estadísticas — El proceso de recopilar, exponer y analizar un conjunto de datos.

Stem-and-Leaf Plot — A plot which uses the digits of the data values to show the shape and distribution of the data set.

```
 5 | 6
 6 |
 7 | 2  5  9  9
 8 | 0  0  6  8  9
 9 | 2  3  4  8
10 | 0  0
Key: 7 | 5 = 75
```

Gráfica de Tallo y Hoja — Un diagrama que utiliza los dígitos de los valores de datos para mostrar la forma y la distribución del conjunto de datos.

```
 5 | 6
 6 |
 7 | 2  5  9  9
 8 | 0  0  6  8  9
 9 | 2  3  4  8
10 | 0  0
Key: 7 | 5 = 75
```

Straight Angle — An angle that measures 180°.

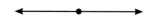

Ángulo Recto — Un ángulo que mide 180°.

English		Spanish	
Straight Edge	A ruler-like tool with no markings.	Borde Recto	Un gobernante como herramienta sin marcas.
Substitution Method	A method for solving a system of linear equations.	Método de Substitución	Un método para resolver un sistema de ecuaciones lineales.
Supplementary Angles	Two angles whose sum is 180°.	Ángulos Suplementarios	Dos ángulos cuya suma es 180°.
Surface Area	The sum of the areas of all the surfaces on a solid.	Área de la Superficie	La suma de las áreas de todas las superficies en un sólido.
System of Linear Equations	Two or more linear equations.	Sistema de Ecuaciones Lineales	Dos o más ecuaciones lineales.

T

Term	A number or the product of a number and a variable in an algebraic expression; a number in a sequence.	Término	Un número o el producto de un número y una variable en una expresión algebraica; un número en una sucesión.
Terminating Decimal	A decimal that stops.	Decimal Terminado	Un decimal que para.
Theorem	A relationship in mathematics that has been proven.	Teorema	Una relación en las matemáticas que ha sido probada.
Theoretical Probability	The ratio of favorable outcomes to the number of possible outcomes.	Probabilidad Teórica	La proporción de resultados favorables a la cantidad de resultados posibles.
Third Quartile (Q3)	The median of the upper half of a data set.	Tercer Cuartil (Q3)	Mediana de la parte superior de un conjunto de datos.
Tick Marks	Equally divided spaces marked with a small line between every inch or centimeter on a ruler.	Marcas de Graduación	Espacios divididos igualmente marcados con una línea pequeña entre cada pulgada o centímetro en una regla.
Transformation	The movement of a figure on a graph so that it changes size or position.	Transformación	El movimiento de una figura en un gráfico de modo que cambia el tamaño o posición

Translation	A transformation in which a figure is shifted up, down, left or right.	Traducción	Una transformación donde la figura se mudo arriba, abajo, a la izquierda o a la derecha.
	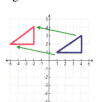		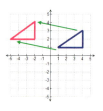
Transversal	A line that intersects two or more lines in the same plane.	Transversal	Una recta que interseca dos o más rectas en el mismo plano.
Trapezoid	A quadrilateral with exactly one pair of parallel sides.	Trapezoide	Un cuadrilateral con exactamente un par de lados paralelos.
	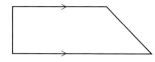		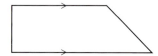
Tree Diagram	A display that organizes information to determine possible outcomes.	Diagrama de Árbol	Una pantalla que organiza la información para determinar los posibles resulatados.
	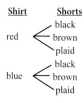		
Trial	A single act of performing an experiment.	Prueba	Un solo intento de realizar un experimento.
Trinomial	An expression with three terms (i.e. $x^2 - 3x + 4$).	Trinomio	Una expreción que tiene tres terminos (es decir: $x^2 - 3x + 4$).
Two-Step Equation	An equation that has two different operations.	Ecuación de Dos Pasos	Una ecuación que tiene dos operaciones diferentes.
Two-Variable Data	A data set where two groups of numbers are looked at simultaneously.	Datos de dos Variables	Un conjunto de datos dónde dos grupos de números se observan simultáneamente.

| Two-Way Frequency Table | A table that shows how many times a value occurs for a pair of categorical data. | Tabla de Frecuencia Bidireccional | Una tabla que muestra cuántas veces aparece un valor de un par de datos categóricos. |

Dog Owners / Walk

Dog Owners	Walk Yes	Walk No
Yes	15	20
No	25	20

Perro Propietario / Paseo

Perro Propietario	Paseo Si	Paseo No
Si	15	20
No	25	20

U-V-W

| Unit Rate | A rate with a denominator of 1. | Índice de Unidad | Un índice con un denominador de 1. |

| Univariate Data | Data that describes one variable (i.e., scores on a test). | Data Univariados | Datos que describen una variable (es decir: puntajes en una prueba). |

| Variable | A symbol that represents one or more numbers. | Variable | Un símbolo que representa uno o más números. |

| Vertex | The minimum or maximum point on a parabola. | Vértice | El mínimo o máximo punto en una parábola. |

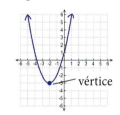

vertex

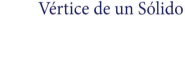

vértice

| Vertex of a Solid | The point where three or more edges meet. | Vértice de un Sólido | El punto donde tres o más bordes se encuentran. |

vertex

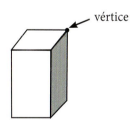

vértice

| Vertex of a Triangle | A point where two sides of a triangle meet. | Vértice de un Triángulo | Un punto donde dos lados de un triángulo se encuentran. |

vertex

vértice

Vertex of an Angle	The common endpoint of the two rays that form an angle. 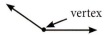	Vértice de un Ángulo	El punto final en común de los dos rayos que forma un ángulo. 
Vertex Form	A quadratic function is in vertex form when written $f(x) = a(x - h)^2 + k$ where $a \neq 0$.	Forma De Vértice	Una función cuadrática es en forma general cuándo escrito $f(x) = a(x - h)^2 + k$ donde $a \neq 0$.
Vertical Angles	Non-adjacent angles with a common vertex formed by two intersecting lines. 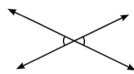	Ángulos Verticales	Ángulos no adyacentes con un vértice en común formado por dos rectas intersecantes. 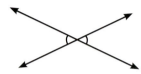
Vertical Line Test	A test used to determine if a graph represents a function by checking to see if a vertical line passes through no more than one point of the graph of a relation.	Examen Vertical De Línia	Un examen para determinar si una gráfica representa una función. Es utilizada para ver si una línia vertical que pasa a través de no más de un punto de la gráfica de una relación.
Volume	The number of cubic units needed to fill a three-dimensional figure.	Volumen	La cantidad de unidades cúbicas necesitadas para llenar un sólido.

X-Y-Z

x-Axis	The horizontal number line on a coordinate plane. 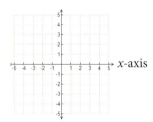	Eje-*x*, Eje de la *x*	La recta numérica horizontal en un plano de coordenadas. 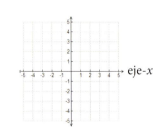

y-Axis	The vertical number line on a coordinate plane.	Eje-*y*, Eje de la *y*	La recta numérica vertical en un plano de coordenadas.

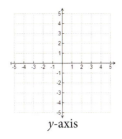

y-axis

eje-*y*

y-Intercept	The point where a graph intersects the *y*-axis.	Intersección *y*	El punto donde una gráfica interseca el eje-*y*.

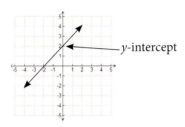

y-intercept

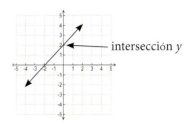

intersección *y*

Zero Pair	One positive integer chip paired with one negative integer chip.	Par Cero	Un chip entero positivo emparejado con un chip entero negativo.

$$1 + (-1) = 0$$

$$1 + (-1) = 0$$

Zero Product Property	If a product of two factors is equal to zero, then one or both of the factors must be zero.	Propiedad De Producto Cero	Si un producto de dos factores es iqual a cero, uno o ambos de los factores debe ser cero.

Zeros	The *x*-intercepts of a quadratic function.	Ceros	Las intersecciones-*x* de una función cuadratica.

zeros, roots, *x*-intercepts

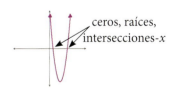

ceros, raíces, intersecciones-*x*

SELECTED ANSWERS

BLOCK 1

Lesson 1.1

1. 17 **3.** 13 **5.** 16 **7.** 71 **9.** 145 **11.** 13 **13.** 50 **15.** 7.6
17. $6 + 3 \times 5 = 21$; see student work **19.** $10 + 4 - 5 \times 2 = 4$ or $10 \times 4 \div 5 \div 2 = 4$; see student work **21.** $10; see student work **23.** Answers may vary.

Lesson 1.2

1. a) 2 **b)** 3 **c)** 8 **3.** 1^4 **5.** 2^7 **7.** 11^3
9. $1 \times 1 \times 1 \times 1 \times 1 \times 1 \times 1 = 1$ **11.** $5 \times 5 \times 5 = 125$
13. $10 \times 10 = 100$ **15.** $\frac{1}{3} \times \frac{1}{3} \times \frac{1}{3} = \frac{1}{27}$ **17.** no; 5 cubed is 125 while 3 to the 5th power is 243 **19.** 2^5 **21.** $4^2, 3^3, 2^5$; see student work **23. a)** $4 \times 4 \times 4$ **b)** 4^3 **c)** 64 cubic units
25. a)

Months Chores Have Been Completed	Pay for Completing Chores	Expanded Form	Power
1	$4		
2	$16	4×4	4^2
3	$64	4×4×4	4^3
4	$256	4×4×4×4	4^4
5	$1024	4×4×4×4×4	4^5

b) Answers may vary.

27. Michelle is correct; see student work **29.** 3 **31.** 125
33. 8 **35.** 1 **37.** $27 \div 9 - 3 = 0$; see student work

Lesson 1.3

1. 35 **3.** 24 **5.** 4 **7.** 112.5 **9.** 39 **11.** 19 **13.** 250 **15.** 69
17. $10 - 4 \div 2^2 = 9$; see student work **19. a)** ants **b)** 409
21. disagree; see student work **23.** 2^5 **25.** 12^6 **27.** 5^3
29. $3^3, 6^2, 2^6$

Lesson 1.4

1. a) Find the value of expressions inside grouping symbols **b)** Find the value of all powers **c)** Multiply and divide from left to right **d)** Add and subtract from left to right **3.** It is important so that all people will get the same answer **5.** 28
7. 27 **9.** 4 **11.** 35 **13.** 39 **15.** 39 **17.** 15
19. $2 \times (4 + 6) = 20$ **21.** $(15 + 1) \times 4 + 5 \times 2 = 74$
23. $\frac{14 + 3 + 4}{3} = 7$; see student work **25.** $3 \times 3 \times 3 = 27$
27. $2 \times 2 \times 2 \times 2 = 16$ **29.** $1 \times 1 \times 1 \times 1 \times 1 = 1$

Lesson 1.5

1. Answers may vary. **3.** Answers may vary. **5.** Associative
7. Commutative **9.** not equal **11.** true; Commutative
13. Answers may vary. **15. a)** 15 and 35 **b)** $15 + 35 + 27$
c) 77 **17.** 24 **19.** 76 **21.** 18 **23.** 0 **25.** 46

Block 1 Review

1. 31 **3.** 43 **5.** 8 **7.** 4^3 **9.** 3^2 **11.** 9^5 **13.** $4 \times 4 \times 4 = 64$
15. $11 \times 11 = 121$ **17.** 2^6 **19. a)** $5 \times 5 \times 5$ **b)** 5^3
c) 125 cubic units **21.** 15 **23.** 100 **25.** 65 **27.** 3 **29.** 94
31. 10 **33. a)** $3 \times (5 + 8) = 39$; see student work
b) $(5 + 3) \times 4 + 12 \div 4 = 35$; see student work
35. Commutative **37.** Associative **39.** true; Associative
41. not equal **43.** $(13 + 6) + 3$

BLOCK 2

Lesson 2.1

1.

Keith's Allowance	Calculation	Michael's Allowance
$14	14 - 3	$11
$20	20 - 3	$17
$25	25 - 3	$22
$31	31 - 3	$28
x	x - 3	x - 3

3. $y + 11$ **5.** $7z$ **7.** $x + 5$ **9.** $w - 31$ **11.** $n - 13$

13. y minus 2 **15.** sixty minus x **17.** r divided by 5
19. x plus 5; 5 more than x; the sum of 5 and x
21. a) 114 beats **b)** 380 beats **c)** $38b$ beats
23. a) $2.00 per pound **b)** $x + 1.25$ **c)** $y - 1.25$
25. $1 \times 1 \times 1 \times 1 \times 1 \times 1 = 1$ **27.** $20 \times 20 = 400$
29. $0 \times 0 \times 0 \times 0 \times 0 = 0$ **31.** not equal

Lesson 2.2

1. 11 **3.** 19 **5.** $18\frac{1}{2}$ **7.** 22 **9.** 15 **11. a)** 420 breaths
b) 1260 breaths **c)** $21m$ **13.** 34 **15.** 7 **17.** 92 **19.** Answers may vary. **21.**

x	6x + 3
0	3
$\frac{1}{2}$	6
3	21
8	51
20	123

23. a) $x =$ number of containers of ice cream; $y =$ number of packages of cones **b)** $20; see student work **25.** See student work; answer = 27
27. $4 \cdot (5 + 20) \div 10 = 10$; see student work **29.** 18 **31.** 24

Lesson 2.3

1. a) area = 15 square units; perimeter = 16 units
b) area = 1 square unit; perimeter = 5 units
c) area = 240 square units; perimeter = 68 units
3. 21 square meters **5.** 28 units **7.** 33 square units
9. 62.8 inches **11.** 216 cubic units **13.** 82 square cm
15. $\approx$ 6 times; see student work **17.** 85 in^2; see student work
19. $6\frac{1}{8}$ square feet **21.** 28 **23.** 39 **25.** 45

Lesson 2.4

1. $24 **3.** $378 **5.** 4 years; see student work **7.** $d = 40$ miles
9. 13.5 miles **11. a)** 558 miles **b)** 2188.5 miles; see student work **13.** 0.284 **15.** 0.290 **17.** Answers may vary. **19.** 4
21. 30 **23.** 1 **25.** 54 square units **27.** 70 square feet

Lesson 2.5

1. $9y$ **3.** $5m$ **5.** 9 **7.** $8u + 8t$ **9.** $11n$ **11.** $3f + 19$
13. $6g + 6k + 10h$ **15.** $b + 4c$ **17.** Answers may vary.
19. She is not correct. She forgot to include the "x" term. The answer should be $5x + 8$ **21.** 5 **23.** 6 **25.** 25 **27.** 3 **29.** $\frac{5}{6}$

Lesson 2.6

1. 5; 4 **3.** 6; 5(2) **5. a)** Answers may vary. **b)** $6(4 + 7) = \$66$
7. $3(20) + 3(4) = 72$ **9.** $7(6) - 7(0.2) = 40.6$
11. $2(60) + 2(8) = 136$ **13.** $5(6) + 5(0.3) = 31.5$
15. Answers may vary. **17.** $7(100 - 1) = 693$
19. $5(1000 + 3) = 5015$ **21.** $8(300 + 5) = 2440$
23. 72 feet **25.** 189 **27.** \$73.50; see student work **29.** 311
31. 53 **33.** 32 **35.** \$1,200 **37. a)** Answers may vary.
b) J.R. - \$48; Lindsey - \$45; Answers may vary.

Lesson 2.7

1. $5x + 35$ **3.** $6m + 6$ **5.** $12p + 8$ **7.** $60 - 3x$ **9.** $22m - 44$
11. Answers may vary. **13.** $6x + 3$ **15.** $5x - 20$ **17.** $6x + 12$
19. $14x + 11$ **21.** $x + 3 + x + 3 + x + 3 + x + 3 + x + 3 = 5x + 15$ or $5(x + 3) = 5x + 15$ **23.** $4x + 1 + x + 2 + 4x + 1 + x + 2 = 10x + 6$ or $2(4x + 1) + 2(x + 2) = 10x + 6$
25. $11(4x - 7) = 44x - 77$ **27.** i. $6x + 3$ ii. $6x + 6$ iii. $6x + 3$; i and iii are equivalent. **29.** $7x + 20$ **31.** $x + 12$ **33.** $7x$
35. #31 → 37, #32 → 150, #33 → 175, #34 → 5

Block 2 Review

1. $d + 9$ **3.** $w - 40$ **5.** $y - 10$ **7.** Answers may vary; six times a number y **9.** Answers may vary; twelve divided by x
11. a) \$2.00 **b)** $y - 0.75$; see student work **13.** 22 **15.** 2
17. 1 **19.**

x	$\frac{1}{2}x - 1$
2	0
6	2
9	$3\frac{1}{2}$
14	6

21. 20 square inches **23.** 38
25. 25.12 units **27.** 46 square cm
29. \$6 **31.** 40 miles **33.** 0.318
35. $15x$ **37.** $10m$ **39.** $9y$
41. $3(20) + 3(2) = 66$
43. $4(5) - 4(0.1) = 19.6$ **45.** $5(200) + 5(1) = 1005$
47. $4(2000) + 4(5) = 8020$ **49. a)** $8(2) - 8(0.03) = 16 - 0.24$
b) \$15.76 **51.** $4x + 40$ **53.** $10x + 6$ **55.** $6x + 17$ **57.** i. $5x + 14$ ii. $5x + 15$ iii. $5x + 15$; ii. and iii. are equivalent

BLOCK 3

Lesson 3.1

1. no **3.** yes **5.** yes **7.** no
9.

x	$9 - x = 3$	Solution?
3	9 - 3 ≠ 3	No
6	9 - 6 = 3	Yes
9	9 - 9 ≠ 3	No

11.

x	$4x = 4.8$	Solution?
1	4(1) ≠ 4.8	No
1.2	4(1.2) = 4.8	Yes
1.8	4(1.8) ≠ 4.8	No

13. a) #2 **b)** 20% **c)** no; answers may vary **15.** $x + 6 = 10$; E
17. $3x - 27$; C **19.** $x + 11 = 14$; D **21. a)** $175 = 7m$ **b)** no
c) more; $7(20) = 140$ is less than 175 **23.** Answers may vary
25. $6x - 16$ **27.** $20x + 35$ **29.** $11x + 2$ **31.** 48 **33.** 10 **35.** 12

Lesson 3.2

1. $x = 6$ **3.** $h = 13$ **5.** $m = 2$ **7.** $y = 23$ **9.** $b = 0$ **11.** $c = 9$
13. a) $c + 98 = 112$ **b)** $c = 14$ pounds **15.** $x + 6 = 15$; $x = 9$
17. $\frac{x}{2} = 7$; $x = 14$ **19.** 235 feet; see student work **21.** $y \approx 9$
23. $x \approx 13$ **25.** $p \approx 7$ **27.** $k = 0$ **29. a)** $s \approx 54$ **b)** The answer represents the total amount spent on souvenirs.
31. 31 tow trucks; see student work for explanation
33.

x	$9x = 45$	Solution?
4	9(4) ≠ 45	No
5	9(5) = 45	Yes
6	9(6) ≠ 45	No

Lesson 3.3

1. $x = 7$ **3.** $p = 8$ **5.** $k = 64$ **7.** $y = 106$ **9.** $f = 1.2$ **11.** $p = 2\frac{1}{2}$
13. a) $x = 42$ **b)** $x = 0$ **c)** Correct **d)** Correct **15.** $x + 12 = 26$; $x = 14$ **17.** $8 + x = 31$; $x = 23$ **19. a)** The variable represents the amount Jared pays for his rent. **b)** \$585
21. $105 + x = 730$; $x = 625$ miles **23.** 19 **25.** 15 **27.** 9
29. 0.15 miles

Lesson 3.4

1. $y = 21$ **3.** $t = 38$ **5.** $k = 40$ **7.** $d = 578$ **9.** $f = 71$ **11.** $f = 6$
13. a) Because she spent (−) \$32 and had \$15 left. **b)** $y = \$47$
c) How much money Carrie had to begin with.
15. a) $t - 9 = 16$ **b)** $t = 25$ turkeys **c)** 5 turkeys per hour
17. $x - 10 = 72$; $x = 82$ **19.** $x - \frac{3}{4} = \frac{1}{2}$; $x = 1\frac{1}{4}$
21. Answers may vary. **23.** 132 square feet
25. 12.56 square feet

Lesson 3.5

1. multiplication **3.** division **5.** $p = 12$ **7.** $h = 80$ **9.** $p = 7$
11. $b = 44$ **13.** $m = 4$ **15.** $x = 8$ **17.** $\frac{x}{4} = 8$; $x = 32$
19. $9x = 36$; $x = 4$ **21.** $5x = 15$; Hannah's sister is 3 years old.
23. \$226,085; see student work for explanation **25.** D **27.** A

Lesson 3.6

1. division **3.** subtraction **5.** subtraction **7.** $x = 31$
9. $p = 7.9$ **11.** $k = 27$ **13.** $m = 101$ **15.** $x = 90$ **17.** $y = 11.24$
19. $a = 0$ **21.** $x = 8$ **23.** $y = 14$ **25.** $y + 8 = 20$; $y = 12$
27. $15x = 45$; $x = 3$ **29.** $\frac{x}{2} = 50$; $x = 100$ **31. a)** $3.2h = 31.20$
b) Horatio's hourly wage is \$9.75. **33. a)** $p - 16 = 19$ **b)** 35 points **35.** $11x$ **37.** $6y$ **39.** $B = 0.220$

Lesson 3.7

1. The length of the rectangle is 5 units. **3.** The width of the cereal box is 3 inches. **5.** The length of the base of the triangle is 8 centimeters. **7.** Emma was walking at a rate of $2\frac{1}{2}$ miles per hour. **9.** Lisa was getting a rate of 3.5%.
11. The height of the sail was 32 feet. **13.** See student work; base = 8 ft **15.** $x = 11$ **17.** $m = 36$ **19.** $c = 104$ **21.** $d = 1\frac{3}{8}$
23. $w = 5$ **25.** $d = 12$

Lesson 3.8

1. $x = 4$ **3.** $x = 50$ **5.** $y = 2$ **7.** $p = 24$ **9.** $x = 1$ **11.** $w = 0$
13. $x = 11$ **15. a)** did not do the inverse operation; $x = 11$
b) subtracted 15 from the wrong side of the equation; $x = 1\frac{1}{2}$
c) did not do the inverse operation; $x = 192$
17. $\frac{y}{2} + 4 = 10$; $y = 12$ **19.** $\frac{1}{2}p - 3 = 4$; $p = 14$
21. $2x + 15 = 27$; $x = 6$ children
23.

x	$4x - 1$
$\frac{1}{2}$	1
3	11
5	19
12	47

25.

x	$(x + 1)^2$
1	4
2	9
5	36
9	100

Block 3 Review

1. no **3.** yes **5.**

x	$3x = 27$	Solution?
7	$3(7) \neq 27$	No
8	$3(8) \neq 27$	No
9	$3(9) = 27$	Yes

7. $x + 8 = 10$
9. $4x = 24$

11. $x = 3$ **13.** $x = 7$ **15.** $y = 20$ **17. a)** $4m = 20$ **b)** \$5; Each milkshake costs \$5. **19.** $y = 24$ **21.** $a = 0.6$ **23.** $x = 2\frac{1}{4}$
25. $x + 22 = 40$; $x = 18$ **27.** $x = 17$ **29.** $p = 46$ **31.** $p = \frac{1}{2}$
33. $x - 5 = 31$; $x = 36$ **35. a)** $w - 42 = 24$ **b)** Christopher needs to drink 66 ounces of water per day; see student work.
37. $p = 9$ **39.** $x = 70$ **41.** $a = 7.5$ **43.** $\frac{x}{7} = 3$; $x = 21$
45. Wesley ate 3 pieces of candy; see student work
47. addition **49.** multiplication **51.** division **53.** $y = 46$
55. $p = 2$ **57.** $a = 186$ **59.** $y = 4\frac{7}{9}$ **61.** The width of the rectangle is 5 units. **63.** The width of the box is 3 *cm*.
65. The length of the base of the triangle is 14 *cm*; see student work for explanation **67.** $x = 4$ **69.** $a = 60$ **71.** $x = 36$
73. $3x + 4 = 25$; $x = 7$ **75.** $2w + 8 = 10$; $w = 1$ **77.** 21 feet; see student work for explanation

BLOCK 4

Lesson 4.1

1. -8 **3.** 7 **5.** 109 **7.**

9.

11. Answers may vary. **13.** 1 **15.** $2\frac{1}{4}$ **17.** 6.8 **19.** -12
21. -8 **23.** $+2$ **25.** Answers may vary.
27. A = 3, B = -3, C = -1, D = 8, E = -6 **29.** Answers may vary; both are correct as it would depend on if someone was ascending or descending the building. **31.** $x = 33$
33. $k = \frac{5}{6}$ **35.** $y = 9.2$ **37.** width = 8 *in* **39.** 6 meters; see student work for explanation

Lesson 4.2

1. Answers may vary. **3.** > **5.** < **7.** > **9.** $-2, -1, 0, 2, 5$
11. $-12, -9, -5, -3, -1$ **13.** $-21, -17, 0, 17, 21$
15. Answers may vary. **17.** > **19. a)** No; his negative numbers are listed in the wrong order.
b) $-10, -7, -3, -1, 0, 1, 4, 9$ **21.** 12 hits **23.** 5% interest rate

Lesson 4.3

1.

3. $(0, 0)$ **5.** $(8, 3)$ **7.** $(0, -4)$
9. $(3, -7)$ **11.** x-axis **13.** y-axis
15. I **17.** III **19.** No; he moved 3 left and 2 up, he should have gone right 2 and down 3. **21. a)** See student work.
b) 18 units **c)** 20 square units **23.** Answers may vary.
25.

x	$2 + 7x$
0	2
1	9
2	16
3	23
5	37

27. $x = 6$ **29.** $u = 99$ **31.** $z = 18\frac{1}{12}$

Lesson 4.4

1. always true **3.** always true **5. a)** rectangle; 10 units²
b) parallelogram; 12 units²
7. parallelogram, rectangle

9. #6 = 16 units; #7 = 26 units **11.** $(4, 4)$ **13.** $(-2, -3)$; see student work for explanation **15.** 12 units²

17. 12 units²

19. a) 90 meters **b)** 190 meters **c)** 380 meters
21.-25. **27.** 90 **29.** 80 **31.** 7

Lesson 4.5

1.

Input x	Function Rule $y = 4x$	Output y
0	4(0)	0
1	4(1)	4
2	4(2)	8
3	4(3)	12

3.

Input x	Function Rule $y = \frac{1}{2}x$	Output y
1	$\frac{1}{2}(1)$	$\frac{1}{2}$
2	$\frac{1}{2}(2)$	1
3	$\frac{1}{2}(3)$	$1\frac{1}{2}$
4	$\frac{1}{2}(4)$	2

5.

Input x	Function Rule $y = \frac{x}{5}$	Output y
0	$\frac{0}{5}$	0
5	$\frac{5}{5}$	1
20	$\frac{20}{5}$	4

7. Answers may vary. $(0, 8), (3, 5), (5, 3), (8, 0)$
9. a)

Input Hour, x	Function Rule $y = 15x$	Output Miles, y
1	15(1)	15
2	15(2)	30
5	15(5)	75
6	15(6)	90

b) $(1, 15), (2, 30), (5, 75), (6, 90)$ **c, d)** **e)** 180 miles

11. $x = 16$ **13.** $x = 10$ **15.** $x = 3.1$ **17.** $x = \frac{7}{12}$ **19.** $x = 3.5$

Lesson 4.6

1. $y = 5 + 4x$ **3.** $y = 2 + 0.5x$ **5.** $y = 1 + 2x$ **7.** $y = 16 - 4x$

9. a)

Input Weeks, x	Output Total Fires Fought, y
0	12
1	15
2	18
3	21
4	24
5	27

b) $y = 12 + 3x$ **c)** 45 fires; see student work **11.** #1: $y = 4x$, #2: $y = 3 + x$ **13. a)** \$15 **b)** \$10 **c)** $y = 15 + 10x$ **d)** \$105 **15.** $y = 4 + 3x$; see student work for explanation **17.** 5 **19.** 32

21. 1 **23.** < **25.** < **27.** <

Lesson 4.7

1.

Input x	Function Rule $y = 1 + 3x$	Output y	Ordered Pair (x,y)
0	1 + 3(0)	1	(0, 1)
1	1 + 3(1)	4	(1, 4)
2	1 + 3(2)	7	(2, 7)
3	1 + 3(3)	10	(3, 10)

3.

Input x	Function Rule $y = 1.5x$	Output y	Ordered Pair (x,y)
0	1.5(0)	0	(0, 0)
1	1.5(1)	1.5	(1, 1.5)
2	1.5(2)	3	(2, 3)
3	1.5(3)	4.5	(3, 4.5)

5.

Input x	Function Rule $y = 7 - \frac{1}{2}x$	Output y	Ordered Pair (x,y)
0	$7 - \frac{1}{2}(0)$	7	(0, 7)
1	$7 - \frac{1}{2}(1)$	$6\frac{1}{2}$	$(1, 6\frac{1}{2})$
2	$7 - \frac{1}{2}(2)$	6	(2, 6)
3	$7 - \frac{1}{2}(3)$	$5\frac{1}{2}$	$(3, 5\frac{1}{2})$

7. start value = 5; amount of change = +2;

9. start value = 5; amount of change = −1;

11. start value = 4; amount of change = $+\frac{1}{2}$ **13.** **15.** $y = 9 + 3x$

17. The lines are parallel; see student work for explanation
19. $y = 2 + 3x$

Lesson 4.8

1. 11, 13, 15; first term = 3; operation = +2
3. 175, 225, 250; first term = 125; operation = +25
5. 9.0, 7.4, 7.0; first term = 9.0; operation = −0.4
7. a) 3 units **b)** 4 units; 5 units **c)** 6 units; 7 units 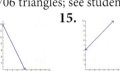 **d)** +1 unit **e)** 11 units; see student work for explanation
9. a) 3 circles; 6 circles; 9 circles **b)** 12 circles; 15 circles **c)** +3 circles **d)** **e)** $y = 3 + 3x$
f) 93 circles; see student work
11. 706 triangles; see student work
13. **15.**

17. **19. a)** $y = 19 - 3x$ **b)** $y = 0.5x$

Lesson 4.9

1. $x > -3$ **3.** $x \geq 1$ **5.** $s \geq 7$;
7. $t < 500$;

9. $m \leq 50$; **11.** B **13.** C
15. C **17.** Answers may vary. > represents greater than while ≥ represents greater than or equal to. **19.** Answers may vary. Wilson is correct. **21.** $y = 12$ **23.** $x = 9$ **25.** $x = 2$
27. $10 + \frac{m}{6} = 13$; $m = 18$ **29.** $4m - 6 = 42$; Pete ran 12 miles; see student work.

Block 4 Review

1. −6 **3.** 11 **5.** 7 **7.** 46 **9.** −5 **11.** < **13.** > **15.** −3, −2, 0, 2, 7
17. a)

Week	Activity	Integer
1	Increased by 5 points	+5
2	Dropped 3 points	−3
3	Gained 10 points	+10
4	Fell 7 points	−7
5	Decreased by 6 points	−6

b) −7, −6, −3, 5, 10

19. (3, 4) **21.** (−4, 2) **23.** (0, 4) **25.** II **27.** y-axis
29. x-axis **31.** Sometimes true. **33.** Sometimes true.
35. parallelogram, rectangle

37. perimeter = 16 units; area = 15 units²

Block 4 Review (Continued)

39.

Input x	Function Rule y = 7x	Output y
0	7(0)	0
1	7(1)	7
2	7(2)	14
3	7(3)	21

41. (2, 4), (4, 10), (5, 13), (8, 22) **43.** $y = 14 - 3x$

45. a) amount of change = −2; It is the coefficient of the variable; **b)** start value = 13 **c)** Decreasing; there is a minus before the variable. **47.** $y = 9 - x$

49.

Input x	Function Rule y = 4 + 2x	Output y	Ordered Pair (x, y)
0	4 + 2 (0)	4	(0, 4)
1	4 + 2 (1)	6	(1, 6)
2	4 + 2 (2)	8	(2, 8)
3	4 + 2 (3)	10	(3, 10)

51.

Input x	Function Rule y = 2.5x	Output y	Ordered Pair (x, y)
0	2.5(0)	0	(0, 0)
1	2.5(1)	2.5	(1, 2.5)
2	2.5(2)	5	(2, 5)
3	2.5(3)	7.5	(3, 7.5)

53. start value = 8; amount of change = −3;

55. $y = 7x$;

57. 31, 39, 47; first term = 7; operation = +8

59. 18, 16, 14, 12, 10; first term = 20; operation = −2

61. $x \geq -4$ **63.** $x > -1$

65. $m \geq 2$

67. $t > 45$

INDEX

PROBLEM-SOLVING

UNDERSTAND THE SITUATION

- ▶ Read then re-read the problem.
- ▶ Identify what the problem is asking you to find.
- ▶ Locate the key information.

PLAN YOUR APPROACH

Choose a strategy to solve the problem:

- ▶ Guess, check and revise
- ▶ Use an equation
- ▶ Use a formula
- ▶ Draw a picture
- ▶ Draw a graph
- ▶ Make a table
- ▶ Make a chart
- ▶ Make a list
- ▶ Look for patterns
- ▶ Compute or simplify

SOLVE THE PROBLEM

- ▶ Use your strategy to solve the problem.
- ▶ Show all work.

ANSWER THE QUESTION

- ▶ State your answer in a complete sentence.
- ▶ Include the appropriate units.

STOP AND THINK

- ▶ Did you answer the question that was asked?
- ▶ Does your answer make sense?
- ▶ Does your answer have the correct units?
- ▶ Look back over your work and correct any mistakes.

DEFEND YOUR ANSWER

Show that your answer is correct by doing one of the following:

- ▶ Use a second strategy to get the same answer.
- ▶ Verify that your first calculations are accurate by repeating your process.

SYMBOLS

Algebra and Number Operations

SYMBOL	MEANING
$+$	Plus or positive
$-$	Minus or negative
$5 \times n, 5 \cdot n, 5n, 5(n)$	Times (multiplication)
$3 \div 4, 4\overline{)3}, \frac{3}{4}$	Divided by (division)
$=$	Is equal to
$\approx$	Is approximately
$<$	Is less than
$>$	Is greater than
$\%$	Percent
$a : b$ or $\frac{a}{b}$	Ratio of a to b
$5.\overline{2}$	Repeating decimal (5.222…)
$\geq$	Is greater than or equal to
$\leq$	Is less than or equal to
x^n	The nth power of x
(a, b)	Ordered pair where a is the x-coordinate and b is the y-coordinate
$\pm$	Plus or minus
$\sqrt{x}$	Square root of x
$\neq$	Not equal to
$x = y$	Is x equal to y?
$\lvert x \rvert$	Absolute value of x
$P(A)$	Probability of event A

Geometry and Measurement

SYMBOL	MEANING
$\cong$	Is congruent to
$\sim$	Is similar to
$\angle$	Angle
$m\angle$	Measure of angle
$\triangle ABC$	Triangle ABC
$\overline{AB}$	Line segment AB
$\overrightarrow{AB}$	Ray AB
AB	Length of AB
π	Pi (approximately $\frac{22}{7}$ or 3.14)
$\circ$	Degree